Moduli Spaces of Polynomials in Two Variables

MEMOIRS
of the
American Mathematical Society

Number 817

Moduli Spaces of Polynomials in Two Variables

Javier Fernández de Bobadilla

January 2005 • Volume 173 • Number 817 (second of 5 numbers) • ISSN 0065-9266

American Mathematical Society
Providence, Rhode Island

2000 *Mathematics Subject Classification.*
Primary 14R05, 14R10, 14H10; Secondary 14E07, 14H20.

Library of Congress Cataloging-in-Publication Data

Fernández de Bobadilla, Javier, 1973–
Moduli spaces of polynomials in two variables / Javier Fernández de Bobadilla.
p. cm. — (Memoirs of the American Mathematical Society, ISSN 0065-9266 ; no. 817)
"Volume 173, number 817 (second of 5 numbers)."
Includes bibliographical references.
ISBN 0-8218-3593-9 (alk. paper)
1. Geometry, Affine. 2. Moduli theory. I. Title. II. Series.
QA3.A57 no. 817
[QA477]
510 s—dc22
[516′.4] 2004057486

Memoirs of the American Mathematical Society

This journal is devoted entirely to research in pure and applied mathematics.

Subscription information. The 2005 subscription begins with volume 173 and consists of six mailings, each containing one or more numbers. Subscription prices for 2005 are $606 list, $485 institutional member. A late charge of 10% of the subscription price will be imposed on orders received from nonmembers after January 1 of the subscription year. Subscribers outside the United States and India must pay a postage surcharge of $31; subscribers in India must pay a postage surcharge of $43. Expedited delivery to destinations in North America $35; elsewhere $130. Each number may be ordered separately; *please specify number* when ordering an individual number. For prices and titles of recently released numbers, see the New Publications sections of the *Notices of the American Mathematical Society.*

Back number information. For back issues see the *AMS Catalog of Publications.*

Subscriptions and orders should be addressed to the American Mathematical Society, P. O. Box 845904, Boston, MA 02284-5904, USA. *All orders must be accompanied by payment.* Other correspondence should be addressed to 201 Charles Street, Providence, RI 02904-2294, USA.

Memoirs of the American Mathematical Society is published bimonthly (each volume consisting usually of more than one number) by the American Mathematical Society at 201 Charles Street, Providence, RI 02904-2294, USA. Periodicals postage paid at Providence, RI. Postmaster: Send address changes to Memoirs, American Mathematical Society, 201 Charles Street, Providence, RI 02904-2294, USA.

This publication is indexed in *Science Citation Index®*, *SciSearch®*, *Research Alert®*, *CompuMath Citation Index®*, *Current Contents®/Physical, Chemical & Earth Sciences.*
Printed in the United States of America.

∞ The paper used in this book is acid-free and falls within the guidelines
established to ensure permanence and durability.
Visit the AMS home page at http://www.ams.org/

10 9 8 7 6 5 4 3 2 1 10 09 08 07 06 05

Contents

Abstract

In the space of polynomials in two variables $\mathbb{C}[x, y]$ with complex coefficients we let the group of automorphisms of the affine plane $\mathbb{A}^2_{\mathbb{C}}$ act by composition on the right. In this paper we investigate the geometry of the orbit space.

We associate a graph with each polynomial in two variables that encodes part of its geometric properties at infinity; we define a partition of $\mathbb{C}[x, y]$ imposing that the polynomials in the same stratum are the polynomials with a fixed associated graph. The graphs associated with polynomials belong to certain class of graphs (called *behaviour graphs*), that has a purely combinatorial definition. We show that any behaviour graph is actually a graph associated with a polynomial. Using this we manage to give a quite precise geometric description of the subsets of the partition.

We associate a moduli functor with each behaviour graph of the class, which assigns to each scheme T the set of families of polynomials with the given graph parametrized over T. Later, using the language of groupoids, we prove that there exists a geometric quotient of the subsets of the partition associated with the given graph by the equivalence relation induced by the action of $Aut(\mathbb{C}^2)$. This geometric quotient is a coarse moduli space for the moduli functor associated with the graph. We also give a geometric description of it based on the combinatorics of the associated graph.

The results presented in this memoir need the development of a certain combinatorial formalism. Using it we are also able to reprove certain known theorems in the subject.

Received by the editor December 5th, 2001.

1991 *Mathematics Subject Classification.* Primary 14R05, 14R10, 14H10. Secondary 14E07, 14H20.

Key words and phrases. moduli spaces, singularities, polynomial maps.

The author was supported by a fellowship from the Banco de España. He wishes to thank for hospitality to the Department of Mathematics of the University of Nijmegen, in which this research was developed.

Introduction

In the recent years the theory of moduli spaces has become very important in Algebraic Geometry. I this realm, when we study a classification problem, we are often interested in the behaviour of families of the objects to be classified, modulo the considered equivalence relation; the construction and study of moduli spaces is usually the most powerful method to study properties of families. Examples of the success of this point of view are the theories of moduli spaces of curves, abelian varieties, vector bundles, stable maps... All of these theories have had striking and often unexpected applications in Algebraic Geometry.

Since the origin of Singularity Theory, one of its main tasks has been the classification of holomorphic germs up to various equivalence relations (e.g. right, right-left and contact equivalence). In the past 20 years there has been a lot of effort in the construction of moduli spaces of germs (see for example [**9**], [**14**] and [**25**]); it has turned out to be a highly non trivial problem that has motivated the development of new tecniques (see for example [**15**]). Only very recently, C. Hertling has been able to construct moduli spaces of hypersurface singularities up to right equivalence in complete generality (see [**19**]); his construction uses an interplay between Frobenius manifolds and Polarized Mixed Hodge Structures in singularities.

In the recent years there is an increasing interest in the study of global polynomial maps; in this new setting, the problem of construction and study of moduli spaces only has been solved in very particular cases. We devote this memoir to the construction and study of moduli spaces of polynomial maps in two variables up to right equivalence. We do not deal here with other equivalence relations, like right-left or contact equivalence, but once our construction is provided one has to do little to obtain moduli spaces in these settings. We have not treated the case of more than two variables, mainly because, in our construction, we need a thorough study of the geometry at infinity of algebraic automorphisms of the plane and affine plane curves, and, at the present moment, such a study seems very complicated in higher dimensions. It is a very interesting (and probably difficult) open problem to try to generalize the results of this memoir to polynomials with more variables.

Consider the space of polynomials in two variables with complex coefficients $\mathbb{C}[x,y]$, and the group $Aut(\mathbb{C}^2)$ of algebraic automorphisms of the complex affine plane $\mathbb{A}^2_{\mathbb{C}}$. The right-equivalence classes are the orbits of the action of $Aut(\mathbb{C}^2)$ on $\mathbb{C}[x,y]$ by composition on the right. The problem of construction and study of moduli spaces that we are considering involves the study of the orbit space $\mathbb{C}[x,y]/Aut(\mathbb{C}^2)$. This quotient is a very complicated space; it is infinite dimensional and non Hausdorff, (even not T_1). A possible approach to understand it better is to define a partition of $\mathbb{C}[x,y]$ such that the quotient of each subset of the partition by the induced equivalence relation exists as an algebraic variety. Then we can study these quotients separately and later study the adjacencies between the subsets of the partition. From this point of view we would have a better (more global) understanding of $\mathbb{C}[x,y]/Aut(\mathbb{C}^2)$ when the partition that we consider is as rough as possible.

In his paper [**29**], M. Oka considers the set $\mathcal{P}ol$, consisting of the polynomials such that their central fibre has only one place at infinity. Given any polynomial $f \in \mathcal{P}ol$, we can associate to it its degree and the tower of weight vectors of a Tschirnhausen-good resolution tower (a special type of toric resolution of the singularity at infinity of f). Oka defines a partition of $\mathcal{P}ol$ by letting each subset of the partition be the set of polynomials of $\mathcal{P}ol$ with given degree and sequence of weight vectors. To fix the degree and the sequence of weight vectors is equivalent to fix the graph of the embedded resolution of the singularity at infinity; thus the subsets of Oka's partition consist of the polynomials with one place at infinity and fixed embedded resolution graph of the singularity at infinity. This partition is such that the quotient of each stratum is a cyclic quotient of a rational variety. In this memoir we improve the results of Oka in the following sense: we define a partition of the whole $\mathbb{C}[x,y]$ (therefore we eliminate the restriction of one place at infinity) such that the quotient of each of the subset is an affine algebraic variety, which, in most of the cases is a quotient of a rational variety by two succesive group actions; the restriction of our partition to $\mathcal{P}ol$ is rougher than Oka's partition.

Although to work with classification problems for polynomials one has to "get the hands dirty" we have tried to take a point of view as geometric and conceptual as possible. The key idea has been to study the interplay between the geometry at infinity of the elements of $Aut(\mathbb{C}^2)$ and the singularities at infinity of the zero set of any polynomial. We associate to each polynomial $f \in \mathbb{C}[x,y]$ a combinatorial property (which we will call *behaviour graph*) that contains less information than the collection of embedded resolution graphs of the singularities at infinity of the locus of vanishing of f. This combinatorial concept enables us to define a partition on $\mathbb{C}[x,y]$ and to construct the desired moduli spaces. In the course of this memoir the interplay between geometry and combinatorics will be constant. We have developed a formalism for the needed combinatorics, which will become central in the proof of many of the main results. Let's summarize briefly the contents of this memoir chapter by chapter.

In the first chapter we analyze the geometry of the elements of $Aut(\mathbb{C}^2)$. For this we study which blowing up processes can arise as resolutions of the indetermination of the extension of an element of $Aut(\mathbb{C}^2)$ to a birational transformation of $\mathbb{P}^2$. As a consequence of we obtain a quite simple geometric-combinatorial new proof of Jung's theorem on factorization of automorphisms (see [**20**]). Our proof is related to the techniques used by Yoshihara in [**38**]. Another proof based on geometric ideas was given by Nagata in [**27**], but it follows different lines.

In the second chapter we introduce the desired partition. For this we associate to any polynomial f a partial resolution of its singularities at infinity, which is, roughly speaking, the partial resolution that consists of the blowing ups of the embedded resolution of the singularities at infinity that can be blowing ups of the resolution of the indeterminacy of an element of $Aut(\mathbb{C}^2)$, viewed as a birational transformation of $\mathbb{P}^2$ (we have cheated a bit, but to give the general idea of how the partition is constructed it is better to do it). To each polynomial together with its partial resolution it is possible to associate a graph (the behaviour graph) in a way that is similar with the classical way of constructing embedded resolution graphs.

We will define the subsets of the partition as the sets of polynomials with the same associated behaviour graph.

In the third chapter we show that each of the subsets of the partition is an algebraic variety. We introduce varieties parametrizing infinitely near points; with the help of them we find a very detailed and surprising description of the geometry of the subsets of the partition: each of them is the total space of a locally trivial fibration over a variety parametrizing infinitely near points whose fibres are complements of a normal crossings arrangement in a complex vector space. Moreover the dimension of the fibres and the number of hyperplanes can be computed from the behaviour graph associated with the subset. For such a geometric description it is essential the use of behaviour graphs instead of embedded resolution graphs of the singularities at infinity.

In the fourth chapter we study the action of $Aut(\mathbb{C}^2)$ in $\mathbb{C}[x,y]$ using as a tool the combinatorics that we will have developed. Our study of the action specializes to a result of W. Neumann and P. Whightwick (see [**28**]) on the change of the degree of polynomials by composition with a sequence of automorphisms. As a consequence we will also deduce the invariance of the behaviour graph by translations on the left (result that is false using embedded resolution graphs of the singularities at infinity instead of the behaviour graph).

In the fifth chapter we formulate in a precise way the moduli problem that we are going to consider. In this formulation we have been influenced by the philosophy of the paper "Quotients by groupoids", by S. Keel and S. Mori (see [**22**]). We use groupoids to reduce our construction of moduli spaces to certain particular cases.

Finally, in the sixth chapter, we construct and study the geometric properties of the quotients of the subsets of the partition by the induced equivalence relation. The way that we construct the moduli spaces is algorithmic: given two polynomials of $\mathbb{C}[x,y]$ we can decide algorithmically whether they are R-equivalent or not. Moreover if they are R-equivalent then it is possible to construct the automorphism that takes one to the other. It is not the first time that such an algorithm appears. In [**28**] W. Neumann and P. Wightwick already presented an algorithm with the same utility. Many of the geometric properties of the moduli spaces can be read directly from their associated graph: the construction of the moduli spaces is done by reduction to the quotienting of a variety by two successive group actions (in most of the cases); many properties of this variety can be read from the graph of the associated graph, and moreover, the groups involved in the quotienting are, the automorphism group of a slight modification of the graph, and a cyclic group, whose order is determined by the graph, respectively.

Apart from the already cited work [**29**],[**28**], and also the paper [**24**], by V.Ya. Lin and M.G. Zaidenberg, I am not aware of other works that deal with the classification of polynomials in two variables *up to right equivalence*. From another point of view D. Siersma and J. Smeltink have worked on the classification of singularities at infinity of polynomials in two variables of degree 4 (see [**32**]). L. Fourrier has worked on the classification modulo topological equivalence at a neighbourhood of infinity (see [**8**]) and, in [**4**], A. Bodin classified (up to topological right equivalence), polynomials with only one atypical value.

I would like to thank M. Guta, D. Siersma, F. Bittner, B. van den Berg, E. Looijenga, J. I. Cogolludo, I. Luengo, A. Melle, and, very specially J. Steenbrink for useful discussions and corrections. I also aknowledge the Department of Mathematics of Nijmegen University for its hospitality and stimulating atmorphere. Finally thanks also to the referee for useful comments.

CHAPTER 1

Automorphisms of the affine plane

Given an automorphism ϕ of the affine plane $\mathbb{C}^2$ we can extend it to a birational transformation of $\mathbb{P}^2$, a completion of $\mathbb{C}^2$ to a projective plane. In this chapter, using basic results of the classical theory of birational transformations of algebraic surfaces, we will study the structure of the indeterminacy locus of ϕ. This study will turn out to be essential for the construction and study of the moduli spaces of polynomials, and also will lead us to a new geometric proof of Jung's theorem on generators of the automorphism group of $\mathbb{C}^2$ (see [**20**]).

1.1. Blowing ups and graphs

In this short section we fix some definitions (most of them classical) on blowing up processes and graphs which we will use extensively along this paper.

DEFINITION 1.1. A blowing up process $\pi = \pi_1 \circ ... \circ \pi_n$ is a sequence of blowing ups at points

$$X_n \xrightarrow{\pi_n} X_{n-1} \xrightarrow{\pi_{n-1}} ...X_1 \xrightarrow{\pi_1} X_0$$

where X_0 is an smooth surface. If L (resp. x) is a smooth curve (resp. a point) in X_0 and the center of π_1 is in L (resp. is x) then we say that π starts in L (resp. x).

NOTATION 1.2. We will denote by E_i the exceptional divisor of π_i. If $X_0 = \mathbb{P}_2$ and π starts at a line L we will denote its strict transform indistinctly by L or by E_0. We use the symbol $\dot{E}$ to denote the difference $E_i \setminus (\cup_{j \neq i} E_j)$.

CONSTRUCTION 1.3. Let π be a blowing up process and let L be a smooth rational curve such that π starts in L. We associate a graph to $\pi^* L$ in the following way: there is a vertex for each irreducible component of $\pi^* L$, weighted with its self-intersection, two vertices are connected with an edge if the components that they represent intersect. Consider the affine plane $\mathbb{C}^2$ and compactify it by fixing an embedding into $\mathbb{P}^2$. When $X_0 = \mathbb{P}^2$ and $L = \mathbb{P}^2 \setminus \mathbb{C}^2$ we can consider two classes of vertices, the one corresponding to L, which will be represented by a white node when we draw the graph, and the others, which will be represented by black nodes.

DEFINITION 1.4. Let $(U, (x, y))$ be a pair formed by an open subset U of an algebraic surface Y and two functions $x, y \in \Gamma(U, \mathcal{O}_Y)$, so that $(x, y) : U \to \mathbb{C}^2$ is an isomorphism. Then we say that $(U, (x, y))$ is an *affine chart* of Y and that (x, y) are the *coordinates* of U; we call *origin* of U the unique point where both x and y vanish.

Consider $\sigma : Y \to U$, the blowing up of U at the origin, then Y can be covered by two affine charts $(V_1, (x_1, y_1))$ and $(V_2, (x_2, y_2))$ characterized by the property that the restriction of σ to V_1 (resp. V_2) is expressed in terms of the coordinates as

$\sigma(x_1, y_1) = (x_1, x_1 y_1)$ (resp. $\sigma(x_2, y_2) = (x_2 y_2, y_2)$). We call to these affine charts *first and second standard blowing up charts* respectively.

Let X be a surface and L a smooth curve in it. Let $\pi : X_n \to X$ be a blowing up process starting in L. Let $(U, (x, y))$ be an affine chart of X_i for a certain i. We will say that $(U, (x, y))$ is an adapted chart to π (or simply an adapted chart, if there is no possible confussion with the blowing up process we are refering to) if and only if the following condition holds:

- If the center of the chart is in $\dot{E}_j$ for any j then $E_j \cap U$ is defined by $y = 0$.
- If the center of the chart is $E_j \cap E_k$ for $j < k$ then $E_j \cap U$ and $E_k \cap U$ are defined respectively by $y = 0$ and $x = 0$.

1.2. Automorphisms of the affine plane

NOTATION 1.5. Throughout this paper $Aut(\mathbb{C}^2)$ will denote the group of algebraic automorphisms of the complex affine plane, $Aut_0(\mathbb{C}^2)$ the subgroup of automorphisms fixing the origin, and G the subgroup of $Aut_0(\mathbb{C}^2)$ consisting of automorphisms whose differential at the origin is the identity. We will usually write H to denote a linear automorphism (i.e an element of $GL(2, \mathbb{C})$) and θ to denote a translation.

DEFINITION 1.6. Let (x, y) be a coordinate system for $\mathbb{C}^2$. We say that:

- $\phi \in Aut(\mathbb{C}^2)$ is a triangular automorphism respect to (x, y) if it is expressed as $\phi(x, y) = (x, y + \sum_{i=2}^{n} a_i x^i)$.
- $\phi \in Aut(\mathbb{C}^2)$ is primitive with respect to (x, y) if it is of the form $\phi = H^{-1} \circ \psi \circ H$, with ψ triangular and $H \in GL(2, \mathbb{C})$

THEOREM 1.7 (Jung [**20**]). *$Aut(\mathbb{C}^2)$ is generated by linear maps, translations and triangular automorphisms.*

A new geometric proof of the previous theorem is given in [**7**]; as the technique and terminology that it involves is central for the construction and study of moduli spaces of polynomials we will reproduce it here. Almost the same proof was obtained independently by S. Lamy (see [**23**]).

Consider the projective plane $\mathbb{P}^2$ together with a projective reference (X, Y, Z). A choice of coordinates (x, y) for $\mathbb{C}^2$ embeds it into $\mathbb{P}^2$ in the following way: declare that the image of the embedding is the open subset defined by $Z \neq 0$ and that $(x, y) = (X/Z, Y/Z)$. Consider $\phi \in Aut(\mathbb{C}^2)$, the choice of a coordinate system (x, y) for $\mathbb{C}^2$ allows us to extend ϕ to a birational transformation of $\mathbb{P}^2$.

DEFINITION 1.8. Consider $\phi \in Aut(\mathbb{C}^2)$ and extend it as a birational transformation of $\mathbb{P}^2$. Let $\pi : X \to \mathbb{P}^2$ be a blowing up process . Denote by ψ the composition $\phi \circ \pi$. Let L the line at infinity of $\mathbb{P}^2$. The points of $\pi^* L$ where ψ is defined are mapped by it to L. An irreducible component E of $\pi^* L$ is called dicritical if the restriction $\psi_{|E} : E \to L$ is dominant.

LEMMA 1.9. *Let ϕ, π and ψ as in definition 1.8. If ψ has no indetermination, then there is a unique dicritical component of $\pi^* L$. If ψ has indetermination, then it has a unique indetermination point and no component of $\pi^* L$ is dicritical.*

PROOF. Let $\sigma : X' \to X$ be the minimal composition of blowing ups, such that $\psi' := \psi \circ \sigma$ has no indetermination. Define $\pi' := \pi \circ \sigma$. If $x \in X$ is an indetermination point of ψ then there exists a component of $\sigma^{-1}(x)$ that is dicritical for ψ'. Otherwise, using Riemann's extension theorem, we could extend ψ to x.

The image by ψ' of the non dicritical components of π' is a finite set Z included in L. The restriction φ of ψ' to $X' \setminus \psi'^{-1}(Z)$ is a finite mapping of degree 1 (its restriction to $\mathbb{C}^2 \subset \mathbb{P}^2 \setminus Z$ is the automorphism ϕ). The cardinality of $\varphi^{-1}(z)$, when $z \in L \setminus Z$, is at least the number of dicritical components of ψ', which is at least the sum of the number of dicritical components plus the number of indeterminacy points of ψ. As the degree of φ is 1 our lemma follows. $\square$

Let's explain how are the blowing up processes that resolve the indeterminacy of primitive automorphisms. Denote by $\mathcal{A}_n$ the graph

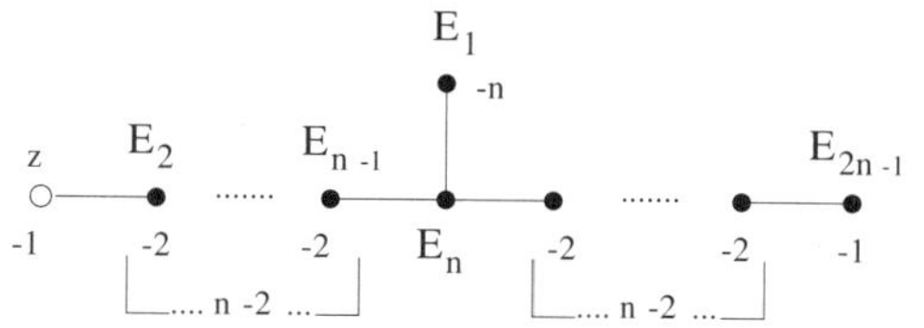

DEFINITION 1.10. We define A_n to be the set of blowing up processes

$$X_{2n-1} \to \ldots \to \mathbb{P}^2$$

starting at the line at infinity whose graph is $\mathcal{A}_n$, and $A'_n \subset A_n$ to be the subset of blowing up processes that start in $(0:1:0)$.

An element in A'_n is determined by the location of the blowing up centers. The first one has to be $(0:1:0)$ by definition. The structure of the graph $\mathcal{A}_n$ implies that:

- for $2 \leq i \leq n$ the blowing up that gives rise to E_i is centered at $E_1 \cap E_i$;
- the i-th blowing up center, for $i \geq n+1$, ranges freely in $\dot{E}_{i-1}$.

In the following picture we see how the exceptional divisors arise in a blowing up process of A'_n. The divisor x corresponds to the strict transform of $X = 0$.

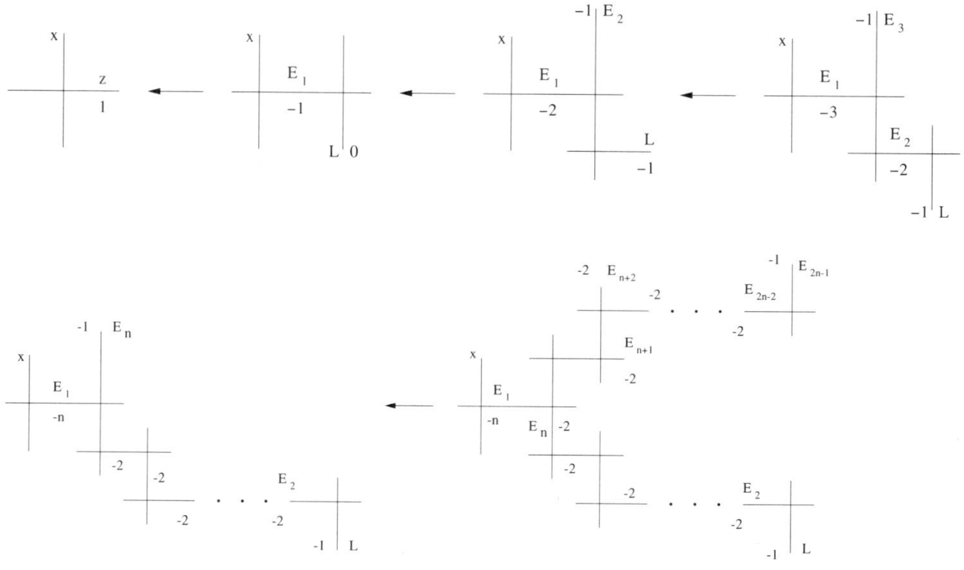

The choice of a projective coordinate system (X, Y, Z) associates to each element of A'_n a birational transformation of $\mathbb{P}^2$ in a unique way: given any

$$\pi : X_{2n-1} \xrightarrow{\pi_{2n-1}} X_{2n-2} \xrightarrow{\pi_{2n-2}} ...X_1 \xrightarrow{\pi_1} \mathbb{P}^2$$

belonging to A'_n, we see that the component L of π^*L has self intersection -1, then, by Castelnuovo's criterion for contractibility we know that there exist a morphism $\pi'_{2n-1} : X_{2n-1} \to X'_{2n-2}$ which contracts L, i.e., a morphism such that X'_{2n-2} is a smooth surface and π'_{2n-1} is the blowing up at a point of X'_{2n-2} whose exceptional divisor is L. The rest of the components of π^*L map by π'_{2n-2} to a set of smooth rational curves with normal crossings. By abuse of language we denote by E_i the image of E_i under π'_{2n-1}. The self-intersection of E_i increases by 1 if E_i was meeting L before the contraction, and remains the same otherwise. The shape of our graph ensures that we can successively contract components of π^*L, finding in each step only one curve with self-intersection -1, until we have contracted all the E_i's except E_{2n-1}, which gets self intersection 1. We have got the following sequence of contractions:

$$\pi' : X_{2n-1} \xrightarrow{\pi'_{2n-1}} X'_{2n-2} \xrightarrow{\pi'_{2n-2}} ...X'_1 \xrightarrow{\pi'_1} X'_0$$

The fact that we contract the same number of curves of X_{2n-1} to get $\mathbb{P}^2$ as to get X'_0 show that the Euler characteristic of X'_0 is the same than the Euler characteristic of $\mathbb{P}^2$; as $\mathbb{P}^2$ is the only complete smooth rational surface with Euler characteristic 3 then X'_0 is isomorphic to it. If we fix an isomorphism $H : X'_0 \to \mathbb{P}^2$ then we can associate to our blowing up process the birational transformation

$$\psi := H \circ \pi' \circ \pi^{-1} \tag{1.1}$$

Hence, to associate to each blowing up process of A'_n a birational transformation from $\mathbb{P}^2$ to $\mathbb{P}^2$ we only need to describe a unique procedure for choosing H, or, what is the same, fixing a projective coordinate system (X', Y', Z') in X'_0. We choose (X', Y', Z') to be the unique projective coordinate system in X'_0 such that the divisor E_{2n+1} is defined by $Z' = 0$ and, if we consider the affine charts $(U_Z, (X/Z, Y/Z))$ and $(U_{Z'}, (X'/Z', Y'/Z'))$ of $\mathbb{P}^2$ and X'_0 respectively, then the restriction $\psi : U_Z \to U_{Z'}$ takes the origin of U_Z to the origin of $U_{Z'}$, having the identity as differential at it. If we restrict the birational transformation ψ to the isomorphism $\psi : U_Z \to U_{Z'}$ and identify each of the affine charts with $\mathbb{C}^2$, we can regard ψ as an element of G.

DEFINITION 1.11. Given a projective coordinate system (X, Y, Z) of $\mathbb{P}^2$, in the above discussion we have defined a mapping

$$t_n : A'_n \to G$$

Define $T_n \subset G$ to be the set of automorphisms of the form $\psi(x, y) = (x, y + \sum_{i=2}^{n} a_i x^i)$, with $a_n \in \mathbb{C}^*$ and $a_i \in \mathbb{C}$ for $i < n$.

LEMMA 1.12. *The mapping t_n is a bijection from A'_n to T_n*

PROOF. To prove the lemma we construct the inverse mapping $s_n : T_n \to A'_n$. Consider $\phi \in T_n$. The choice of coordinates (x, y) in $\mathbb{C}^2$ fixes a unique embedding of $\mathbb{C}^2$ into $\mathbb{P}^2$. Then ϕ extends to a birational transformation of $\mathbb{P}^2$, whose unique indetermination point is $(0 : 1 : 0)$. Let $\pi : X \to \mathbb{P}^2$ be the minimal blowing up process that resolves the indetermination of ϕ. It is direct to check that the graph associated with π is $\mathcal{A}_n$. As the first blowing up of π is centered at $(0:1:0)$ we conclude that $\pi \in A'_n$. Define $s_n(\phi) := \pi$.

Define $\psi := \phi \circ \pi$. From the theory of birational morphisms of smooth surfaces we conclude that ψ is the successive contraction of $L, E_2, ..., E_{2n-2}, E_1$. If we are given the blowing up process π and we want to construct $t_n(\pi)$, then the choice of the coordinate system (X', Y', Z') in the target $\mathbb{P}^2$ (that here plays the role of X'_0) should be equal to the coordinate system (X, Y, Z): with this choice $\psi \circ \pi^{-1}$ is expressed with respect to the coordinates of the affine charts $(U_Z, (X/Z, Y/Z))$ and $(U_{Z'}, (X'/Z', Y'/Z')) = (U_Z, (X/Z, Y/Z))$ as $\psi(x,y) = (x, y + \sum_{i=2}^n a_i x^i)$ and hence it takes the origin to the origin and the differential at the origin is the identity. This shows that $t_n \circ s_n = id_{T_n}$.

To finish the proof we only need to show that s_n is surjective. The mapping s_n can be extended to a mapping $s'_n : t_n(A'_n) \to A'_n$ by defining $s_n(\phi)$ to be the minimal blowing up process that resolves the indetermination of the extension of ϕ to the projective plane. It is obvious that $s'_n \circ t_n = id_{A'_n}$ and hence that t_n is injective. We would finish the proof if we show that $T_n = t_n(A'_n)$.

Consider the family of automorphisms $h_\lambda(x,y) = (x, y + \lambda x)$. Note that any element of T_n commutes with any automorphism of the family. We claim that any automorphism in G that commutes with all the automorphisms of the family is triangular. Let $\phi(x,y) = (f(x,y), g(x,y))$ be an automorphism of G that commutes with the whole family. Then $f(x,y) = f(x, y + \lambda x)$ for any $\lambda \in \mathbb{C}$, and hence f should be a polynomial involving only the variable x. Using that the jacobian of any automorphism must be a unit, an easy computation shows that ϕ must be triangular.

Next we show that any automorphism in $t_n(A'_n)$ commutes with h_λ for any $\lambda \in \mathbb{C}$. Consider $\phi \in t_n(A'_n)$, to show that it commutes with h_λ we will show that $\phi' := h_\lambda \circ \phi \circ h_\lambda^{-1}$ is equal to ϕ. The automorphism h_λ can be extended to the projective automorphism $H_\lambda(X:Y:Z) := (X, Y + \lambda X : Z)$ and then the birational transformation $\phi' : \mathbb{P}^2 \dashrightarrow \mathbb{P}^2$ is equal to $H_\lambda \circ \phi \circ H_\lambda^{-1}$. We show that the blowing up processes needed to resolve the indeterminacy of ϕ and ϕ' are the same: let

$$\pi : X_{2n-1} \xrightarrow{\pi_{2n-1}} X_{2n-2} \to ... \to X_1 \xrightarrow{\pi_1} \mathbb{P}^2$$

be the resolution of the indeterminacy of ϕ. Denote by x_i the blowing up center of π_i. We know that $x_1 = (0:1:0)$. Let $\psi_i := \phi \circ \pi_1 \circ ... \pi_i$, then x_i is the unique indetermination point of ψ_{i-1}. As ϕ' is the conjugation of ϕ by H_λ we deduce that the unique indetermination point of ϕ' is $H_\lambda(0:1:0) = (0:1:0)$. Let $\psi'_1 := \phi' \circ \pi_1$. The isomorphism H_λ lifts to an isomorphism $A_1 : X_1 \to X_1$ and ψ'_1 is the conjugation of ψ_1 by A_1. Hence the indetermination point x'_2 of ψ'_1 equals $A_1(x_2)$. As $x_1 = E_0 \cap E_1$, a necessarily fixed point of A_1, then $x'_2 = x_2$. We can proceed in the same way and show that the n first blowing ups centers of both resolutions of indeterminancy coincide. Let $\psi'_n := \phi' \circ \pi_n \circ ... \pi_1$. The isomorphism H_λ lifts to an isomorphism $A_n : X_n \to X_n$ and the indetermination point x'_{n+1} of ψ'_n is $A_n(x_{n+1})$. An easy computation shows that if we take the standard blowing up chart $(V, (u,v))$ centered in $E_1 \cap E_n$, for which E_n is defined by $v = 0$ then A_n has the following expression respect (u,v):

$$A_n(u,v) = (\frac{1}{1 + \lambda u v^{n-1}} u, v) \tag{1.2}$$

and hence $A_n|E_r = id_{E_r}$. As $x_{n+1} \in \dot{E}_n$ we have that $x'_{n+1} = A_n(x_{n+1}) = x_{n+1}$. Define $\psi'_{n+1} := \phi' \circ \pi_{n+1} \circ ... \pi_1$. Let A_{n+1} be the lifting of A_n to X_{n+1}. The indetermination point x'_{n+2} of ψ'_{n+1} equals $A_{n+1}(x_{n+2})$. The point x_{n+2} is in $\dot{E}_{n+1}$. A computation shows that A_{n+1} fixes the points of E_{n+1}.

In an analogous way we can show that $x_i = x'_i$ for any $i \leq 2n-1$. As $\psi'_{2n-1} := \phi' \circ \pi_{2n-1} \circ ... \circ \pi_1$ is the conjugation of the morphism ψ_{2n-1} by the lifting of H_λ to X_{2n-1} we deduce that ψ'_{2n-1} has no indetermination. We conclude that both ϕ and ϕ' need the same blowing up process π to resolve their indeterminacy. The same argument that has been used to show that $t_n \circ s_n = id_{T_n}$ shows that $\phi' = t_n(\pi)$. As, $\phi = t_n(\pi)$, because of the injectivity of t_n we deduce that $\phi = \phi'$. This shows the commutativity.

We deduce that any $\phi \in t_n(A'_n)$ must be triangular. If $\phi \in T_m$ for $m \neq n$ then the graph of the resolution of its indeterminacy would be $\mathcal{A}_m$ instead of $\mathcal{A}_n$. This concludes the proof. □

REMARK 1.13. We can see s_n as a mapping that assigns to each triangular automorphism $\psi \in T_n$ two blowing up processes $\pi : X \to \mathbb{P}^2$, $\pi' : X \to \mathbb{P}^2$ of A'_n such that if we understand ψ as a birational transformation from $\mathbb{P}^2$ to $\mathbb{P}^2$ then $\psi := \pi' \circ \pi^{-1}$

PROOF. Just let π be as in the proof above and $\pi' = \psi_{2n+1}$. □

The following picture represents the decomposition in blowing ups and blowing downs of the automorphism $\psi(x, y) = (x, y - x^2)$. The coordinates of the source $\mathbb{C}^2$ are called x, y The coordinates of the target $\mathbb{C}^2$ are x', y'.

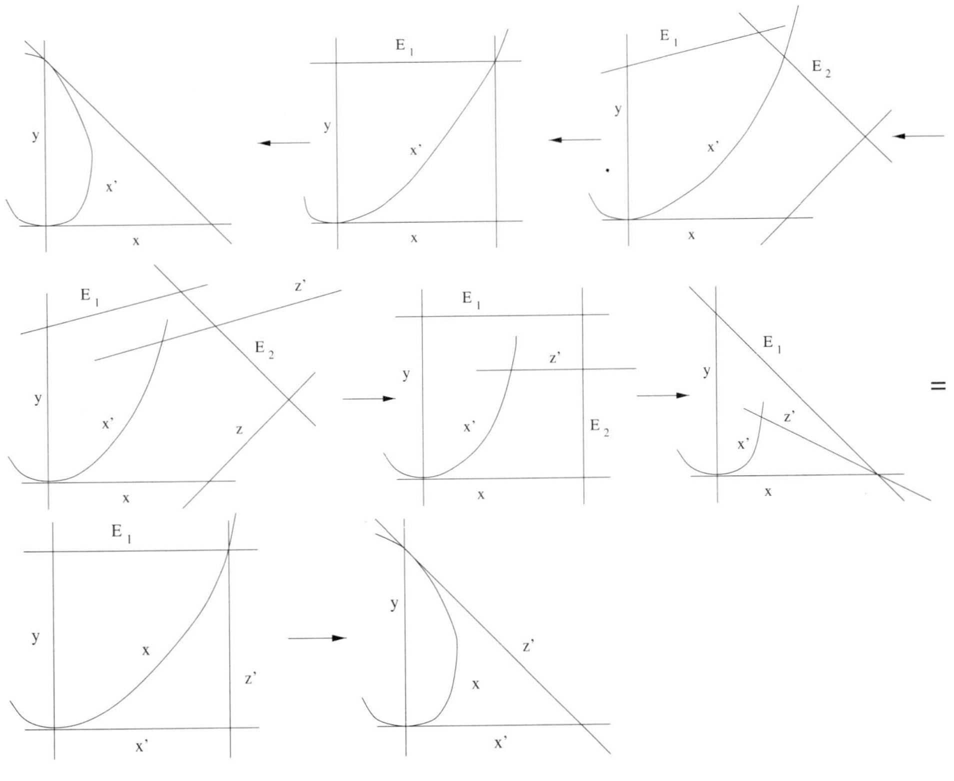

DEFINITION 1.14. The discussion that preceeds definition 1.11 gives also a mapping

$$t_n : A_n \to G$$

Define P_n to be the set of primitive automorphisms that are conjugate to a triangular automorphism of T_n.

LEMMA 1.15. *t_n defines a bijection from A_n onto P_n.*

PROOF. Consider a blowing up process π of A_n. Let the center of the first blow up be $p \in L$, where L is the line at infinity. Consider $H \in GL(2, \mathbb{C})$ such that $H(p) = (0 : 1 : 0)$. If we change the projective coordinate system to $(H^*X : H^*Y : H^*Z)$ then the coordinates of p become $(0 : 1 : 0)$ and our blowing up process becomes an element of A'_n. Denote by t'_n the mapping analogous to t_n but for the coordinate system $(H^*X : H^*Y : H^*Z)$. By Lemma 2, $t'_n(\pi)$ is an element of T_n (respect to the coordinate system $(H^*X : H^*Y : H^*Z)$. We have that $t_n(\pi) = H^{-1} \circ t'(\pi) \circ H$ because they have the same minimal resolution of indeterminancy, both of them send the origin of $\mathbb{C}^2$ to the origin of $\mathbb{C}^2$ with differential at it equal to the identity; therefore $t_n(\pi)$ belongs to P_n. Hence $t(A_n) \subset P_n$. Bijectivity follows easily. □

REMARK 1.16. If $s_n : P_n \to A_n$ is the inverse of t_n then, given any $\phi \in P_n$, the blowing up process $s_n(\phi)$ is the minimal resolution of indeterminancy of ϕ. We can see s_n as a mapping that assigns to each primitive automorphism $\psi \in P_n$ two blowing up processes $\pi : X \to \mathbb{P}^2$, $\pi' : X \to \mathbb{P}^2$ of A_n such that if we understand ψ as a birational transformation of $\mathbb{P}^2$ to then $\psi := \pi' \circ \pi$.

PROOF OF THEOREM 1.7. Consider any $\phi \in Aut(\mathbb{C}^2)$. Let $\pi : \pi_1 \circ ... \circ \pi_n$ be the minimal resolution of the indetermination of ϕ as a birational transformation of $\mathbb{P}^2$. Let E_i be the exceptional divisor of π_i, define $\sigma_i := \pi_1 \circ ... \circ \pi_i$. Lemma 1.9 implies that π_{i+1} is the blowing up at the unique indetermination point x_{i+1} of $\phi \circ \sigma_i$, for any $i \leq n-1$, and that the unique dicritical component of $\pi^* L$ is the last exceptional divisor E_n. Moreover, x_i belongs to E_{i-1} if $i \geq 2$ because, otherwise, $\phi \circ \pi_{i-2}$ would have two indetermination points. The theory of birational transformations of smooth surfaces implies that $\phi \circ \pi$ equals $H \circ \pi'$ where $H : Y \to \mathbb{P}^2$ is an isomorphism and $\pi' : X \to Y$ is the successive contraction of the non dicritical components of $\pi^* L$ with self-intersection -1. We claim that there exists r such that the graph of σ_{2r-1} is $\mathcal{A}_r$.

Let E_i be a component of $\pi^* L$ different from L and E_n. It has self-intersection strictly smaller than -1: its initial self intersection is -1, and it decreases by 1 when we blow up at $x_{i+1} \in E_i$. As E_n is the exceptional divisor of the last blowing up it has self-intersection -1. The strict transform of the line at infinity L should have self-intersection -1, otherwise in the contraction process $\phi \circ \pi$ the only possible divisor to start with is E_n, and it is dicritical. Before we start blowing up, L has self-intersection 1; as x_1 meets L the self-intersection of L becomes 0 after π_1; as we have to decrease it to -1 another blowing up center should meet L, the only possible one is x_2 (use that x_i should meet E_{i-1}). After π_2 the self-intersection of L is already -1 and hence no more blowing up centers meet L. The center x_3 can be either $E_1 \cap E_2$ or a point in $\dot{E}_2$. In the last case the claim is true for $r = 2$. Hence we assume that $x_3 = E_1 \cap E_2$. Let r be the maximal number such that

$x_i = E_1 \cap E_{i-1}$ for any $3 \leq i \leq r$. The divisor E_r is non dicritical: otherwise it should be possible to successively contract all the components except E_r starting by L. The self-intersection of E_1 is $-r$ and the divisors L, $E_2...$, E_{r-1} are separated from E_1 by E_r, hence in the contraction process E_1 would never increase its self-intersection, and hence it could never be contracted. We conclude that there is a further blowing up π_{r+1} in the blowing up process π. The center of π_{r+1} should be either $E_{r-1} \cap E_r$ or a point of $\dot{E}_r$.

If $x_{r+1} = E_{r-1} \cap E_r$ then, after π, the self-intersection of E_{r-1} is upper bounded by -3. Remembering that only the non dicriticals with self-intersection -1 can be contracted we easily see that the contraction process π' starts with the succesive contraction of L, E_2,...,E_{r-2}. After this E_{r-1} gets self-intersection upper bounded by -2, as we have contracted only one component that meets it. The self-intersection of the rest of the remaining components is not affected by the contractions. Then the only component with self-intersection -1 is the dicritical component and hence we cannot finish the contraction procedure. We conclude that $x_{r+1} \in \dot{E}_r$.

Let s be the maximal integer such that x_{r+i} belongs to $\dot{E}_{r+i-1}$ for $1 \leq i \leq s$. We prove that $s \geq r-1$; we are dealing with $r \geq 3$. We assume that $s < r-1$. Because of the definition of s we have that either E_{r+s} is dicritical or x_{r+s+1} equals $E_{r+s-1} \cap E_{r+s}$. In both cases the divisor L has self intersection -1, the divisor E_1 has $-r$, the divisors E_i have -2, for $2 \leq i \leq s+r-2$. If E_{r+s} is dicritical then the self-intersection of E_{r+s-1} equals -2. If we contract successively the non dicritical components with self-intersection -1 we will reach a point in which the only remaining components will be E_{r+s}, that is dicritical, and E_1, with self-intersection $-r+s < -1$. Hence the contraction process cannot be completed. If $x_{r+s+1} = E_{r+s-1} \cap E_{r+s}$ then the self-intersection of E_{r+s-1} is strictly smaller than -2. We can contract successively L, E_2,...,E_r,...,E_{r+s-2}. After this E_{r+s-1} has self-intersection strictly smaller than -1, because the only meeting it before being contracted was E_{r+s-2}. The rest of the remainig non dicritical have self intersection upper bounded by -2 because they are separated from the contracted components by E_{r+s-1}. Hence the contraction process again cannot be completed. This proves that $s \geq r-1$, and this, in turn, implies that the graph of σ_{2r-1} is $\mathcal{A}_r$, as we claimed.

Let $\sigma_{2r,n}$ be the composition $\pi_{2r} \circ ... \circ \pi_n$. Define σ'_{2r-1} as the composition of the first $2r-1$ contractions of π' and $\sigma'_{2r,n}$ as the composition of the rest of the contractions. We have that $\phi = H \circ \pi' \circ \pi^{-1}$. If $\sigma_{2r,n}$ is not trivial (when $\pi \neq \sigma_{2r-1}$) then x_{2r} is in $\dot{E}_{2r-1}$, this implies the commutativity $\sigma'_{2r,n} \circ \sigma'_{2r-1} \circ \sigma^{-1}_{2r,n} \circ \sigma^{-1}_{2r-1} = \sigma'_{2r,n} \circ \sigma^{-1}_{2r,n} \circ \sigma'_{2r-1} \circ \sigma^{-1}_{2r-1}$. As the graph of σ_{2r-1} is $\mathcal{A}_r$ there exists a unique isomorphism F from the target of σ'_{2r-1} to $\mathbb{P}^2$ that makes $t_n(\sigma_{2r-1}) = F \circ \sigma'_{2r-1} \circ \sigma^{-1}_{2r-1}$. If we define $\phi'' := H \circ \sigma'_{2r,n} \circ \sigma^{-1}_{2r,n} \circ F^{-1}$ then we have the factorization $\phi = \phi'' \circ t_n(\sigma_{2r-1})$, where $t_n(\sigma_{2r-1})$ is primitive and ϕ'' needs less blowing ups than ϕ to resolve its indetermination. Hence the theorem is proved by induction on the number of blowing ups needed to resolve the indetermination. □

We have the following reformulation of Jung's theorem:

LEMMA 1.17. *Given $\phi \in Aut(\mathbb{C}^2)$ there exists a unique decomposition as $\phi = \theta \circ H \circ \psi_n \circ \ldots \psi_1$, where θ is a translation, $H \in GL(2, \mathbb{C})$ and $\psi_i \in G$ are non trivial* *primitive automorphisms such that the unique indetermination points of* ψ_i *and* ψ_{i+1}*, regarded as birational maps from* $\mathbb{P}^2$ *to* $\mathbb{P}^2$*, are different.*

PROOF. Direct after the last proof. □

DEFINITION 1.18. Denote by $G_{n_1,...,n_r}$ the subset of G consisting of those automorphisms whose decomposition into primitive automorphisms is of the form $\phi = \psi_r \circ ... \circ \psi_1$ with $\psi_i \in P_{n_i}$.

Denote by $\mathcal{A}_{n_1,...,n_r}$ the graph

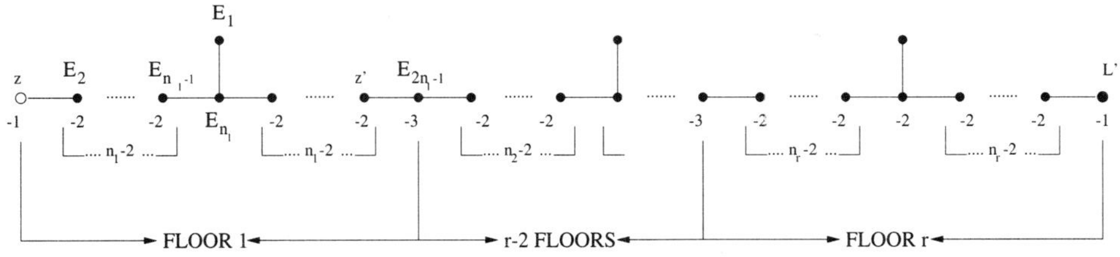

DEFINITION 1.19. Define $A_{n_1,...,n_r}$ to be the set of blowing up processes

$$X_m \to ... \to X_1 \to \mathbb{P}^2$$

starting on L such that their associated graph is $\mathcal{A}_{n_1,...,n_r}$.

We define a map

$$t_{n_1,...,n_r} : A_{n_1,...,n_r} \to G$$

as follows: given a blowing up process $\pi \in A_{n_1,...n_r}$, the shape of $\mathcal{A}_{n_1,...,n_r}$ implies that we can successively contract irreducible components of π^*L with self-intersection -1 starting with the strict transform of the line at infinity, until E_m is the only remaining one that cannot be contracted, because, after the previous contractions, it gets self-intersection 1. Denote by $\pi' : X \to X_0'$ the contraction morphism. An Euler characteristic argument like before shows that X_0' is isomorphic to $\mathbb{P}^2$. In X_0' we consider the unique projective coordinate system $(X', Y,' Z')$ such that $Z' = 0$ defines the locus E_m and such that, given the affine charts $(U_Z, (X/Z, Y/Z))$ and $(U_{Z'}, (X'/Z', Y'/Z'))$ of $\mathbb{P}^2$ and X_0' respectively, the restriction $\pi' \circ \pi^{-1} : U_Z \to U_{Z'}'$ takes the origin to the origin and the differential at the origin is the identity. We define $t_{n_1,...,n_r}(\pi)$ to be the isomorphism of $\mathbb{C}^2$ that $\pi' \circ \pi^{-1}$ gives if we identify each of the affine charts with $\mathbb{C}^2$ via the isomorphisms defined by the coordinate functions.

PROPOSITION 1.20. *The map* $t_{n_1,...,n_r}$ *defined in the last discussion is a bijection from* $A_{n_1,...,n_r}$ *to* $G_{n_1,...,n_r}$.

PROOF. We prove the proposition by defining an inverse

$$s_{n_1,...,n_r} : G_{n_1,...,n_r} \to A_{n_1,...,n_r}$$

Consider $\psi = \psi_r \circ ... \circ \psi_1 \in G_{n_1,...,n_r}$. By Remark 1.16, for each i the mapping s_{n_i} associates to ψ_i two blowing up processes $\pi_i : X_i \to \mathbb{P}^2$ and $\pi_i' : X_i \to \mathbb{P}^2$ such that $\psi_i = \pi_i' \circ \pi_i^{-1}$. This gives a factorization of the form $\psi = \pi_r' \circ \pi_r^{-1} \circ ... \circ \pi_1' \circ \pi_1^{-1}$

We look at each π_i as the composition of

$$X_i =: X_{i,2n_i-1} \stackrel{\pi_{i,2n_i-1}}{\longrightarrow} X_{i,2n_i-2} \to ... \to X_{i,1} \stackrel{\pi_{i,1}}{\longrightarrow} \mathbb{P}^2$$

where $\pi_{i,j}$ is the blowing up of $X_{i,j-1}$ at a point $x_{i,j}$. Let $E_{i,j}$ be the exceptional divisor corresponding to the blowing up at $x_{i,j}$, and denote by L_i the strict transform of the line at infinity under π_i. Look π'_i as the composition of

$$X_i =: X'_{i,0} \xrightarrow{\pi'_{i,0}} X'_{i,1} \xrightarrow{\pi'_{i,2}} X'_{i,2} \to \dots \to X'_{i,2n_i-3} \xrightarrow{\pi'_{i,2n_i-2}} X'_{i,2n_i-2} \xrightarrow{\pi'_{i,1}} \mathbb{P}^2$$

where $\pi'_{i,0}$ is the contraction of L_i and $\pi'_{i,j}$ is the contraction of $E_{i,j}$.

Denote by D'_1 the exceptional divisor of π'_1. As the restriction $\pi'_1 : X_1 \setminus D_1 \to \mathbb{P}^2 \setminus \pi'_1(D_1)$ is an isomorphism and $x_{2,1}$ lies in $\mathbb{P}^2 \setminus \pi'_1(D_1)$ the points $y_{2,j} := \pi_1'^{-1}(x_{2,j})$ for $1 \leq j \leq 2n_2 - 1$ are well defined infinitesimally near points of X_1. Define the sequence of blowing ups

$$Y_{2,2n_2-1} \xrightarrow{p_{2,2n_2-1}} Y_{2,2n_2-2} \to \dots \to Y_{2,1} \xrightarrow{p_{2,1}} Y_{2,0} := X_1$$

where $p_{2,j}$ is the blowing up of $Y_{2,j-1}$ at $y_{2,j}$. Let $E_{2,j}$ be the exceptional divisor of the blowing up $p_{2,j}$ and p_2 the composition of all $p_{2,j}$. By abuse of language, for any $1 \leq j \leq 2n_1 - 1$, denote by $E_{1,j}$ the strict transform of $E_{1,j}$ by p_2, and by L the strict transform of the line at infinity. Define also the sequence

$$Y_{2,2n_2-1} =: Y'_{1,0} \xrightarrow{p'_{1,0}} Y'_{1,1} \xrightarrow{p'_{1,2}} Y'_{1,2} \to \dots \to Y'_{1,2n_1-3} \xrightarrow{p'_{1,2n_1-2}} Y_{1,2n_1-2} \xrightarrow{p'_{1,1}} Y_{1,2n_1-1}$$

where $p'_{1,0}$ is defined to be the contraction of L and $p'_{1,j}$ the contraction of $E_{1,j}$. Denote by p'_1 the composition of all the $p'_{1,j}$'s. Notice that $Y'_{1,2n_1-1}$ can be naturally identified with X_2 in such a way that $\pi_2^{-1} \circ \pi'_1 = p'_1 \circ p_2^{-1}$. Define

$$\pi_{1-2} := \pi_1 \circ p_2 \qquad \pi'_{1-2} := \pi'_2 \circ p'_1 \qquad X_{1-2} := Y_{2,2n_2-1}$$

$\pi_{1-2} : X_{1-2} \to \mathbb{P}^2$ is a blowing up process with graph $\mathcal{A}_{n_1,n_2}$, $\pi'_{1-2} : X_{1-2} \to \mathbb{P}^2$ is a blowing up process with graph $\mathcal{A}_{n_2,n_1}$ and we have expressed $\pi'_2 \circ \pi_2^{-1} \circ \pi'_1 \circ \pi_1^{-1}$ as $\pi'_{1-2} \circ \pi_{1-2}^{-1}$.

By using an analogous reasoning we can find blowing up processes $\pi : X \to \mathbb{P}^2$ with graph in $\mathcal{A}_{n_1,\dots,n_r}$ and $\pi'; X \to \mathbb{P}^2$ with graph $\mathcal{A}_{n_r,\dots,n_1}$ such that

$$\psi = \pi'_r \circ \pi_r^{-1} \circ \dots \circ \pi'_1 \circ \pi_1^{-1} = \pi \circ \pi^{-1}$$

We define $s_{n_1,\dots,n_r}(\psi) := \pi$. Note that π is the minimal blowing up process that resolves the indeterminacy of ψ.

We show that $t_{n_1,\dots,n_r} \circ s_{n_1,\dots,n_r} = id_{G_{n_1,\dots,n_r}}$. Let $\pi : X \to \mathbb{P}^2$ be $s_{n_1,\dots,n_r}(\psi)$ for $\psi \in G_{n_1,\dots,n_r}$. To compute $t_{n_1,\dots,n_r}(\pi)$ we need to specify a contraction process $\pi' : X \to X'_0$ and a projective coordinate system (X', Y', Z') of X'_0. The contraction process π' is exactly the morphism $\pi' : X \to \mathbb{P}^2$ that appears in the contruction of $s_{n_1,\dots,n_r}(\psi)$, and if we take in $X'_0 = \mathbb{P}^2$ the projective coordinate system (X', Y', Z') to be equal to the fixed projective coordinate system (X, Y, Z) then the expression of $\pi' \circ \pi^{-1}$ respect to the coordinates of the affine charts $(U_Z, (X/Z, Y/Z))$ and $U'_Z, (X'/Z', Y'/Z'))$ is the expresion of ψ respect to the coordinates $(x, y) = (X/Z, Y/Z)$. This shows that (X', Y', Z') must be chosen equal to (X, Y, Z) and that $t_{n_1,\dots,n_r} \circ s_{n_1,\dots,n_r}(\psi) = \psi$.

To show that $s_{n_1,\dots,n_r} \circ t_{n_1,\dots,n_r} = id_{A_{n_1,\dots,n_r}}$ we notice that the minimal blowing up process that resolves the indeterminacy of $t_{n_1,,n_r}$ is π. This finishes the proof. □

The arguments of the proof above allow us to interpret geometrically the elements of $Aut(\mathbb{C}^2)$ as follows:

Definition 1.21. Define $Aut(\mathbb{C}^2)_{n_1,\dots,n_r}$ to be the set of automorphisms $\phi \in Aut(\mathbb{C}^2)$ that can be expressed as $\phi = \theta \circ H \circ \psi$ with θ a translation, $H \in GL(2,\mathbb{C})$ and $\psi \in G_{n_1,\dots,n_r}$.

Corollary 1.22. *Given any $\phi \in Aut(\mathbb{C}^2)_{n_1,\dots,n_r}$ there exist a unique factorizarion of the form*

$$\phi = \theta \circ H \circ \pi^{-1} \circ \pi'$$

with θ a translation, $H \in GL(2,\mathbb{C})$, π, π' belonging to $A_{n_1,\dots,n_r}$.

CHAPTER 2

A partition on $\mathbb{C}[x, y]$

The object of this memoir is the construction and geometric study of moduli spaces of polynomials in two variables modulo R-equivalence. The orbit space $\mathbb{C}[X,Y]/Aut(\mathbb{C}^2)$ of polynomials modulo R-equivalence is infinite dimensional and non separated, (even not T_1); therefore it is quite difficult to say something about it studying it as a whole. In order to understand it better we give a geometrically meaningful partition of $\mathbb{C}[X,Y]$ in subsets such that the quotient of each of them by the equivalence relation induced from the action of $Aut(\mathbb{C}^2)$ exists as an algebraic variety. The main idea to define the partition is, given any polynomial, to compare the blowing up process that resolves it singularities at infinity, with each of the blowing up processes that resolve indeterminancy of polynomials.

For the precise definition of the partition we will use a combinatorial formalism that will turn out to be very convenient for the construction and geometric study of polynomials.

2.1. Parental systems and graphs

In this section we define a class of graphs needed for the definition and study of the partition on $\mathbb{C}[x, y]$. They are quite similar to the graphs used in the literature to represent embedded resolutions of plane curve singularities. The formalism that we use to define them is new and will help in making easier to formulate the results and proofs of this paper. The best way of introducing our class of graphs is via the following definition, which follows a suggestion of Steenbrink:

Let S be a set and w a distinguished element of it. Let $\binom{S}{2}$ be the set whose elements are the subsets of S of cardinality two. Identify S with the set of its subsets of cardinality 1. Consider a mapping

$$p : S \setminus \{w\} \to S \coprod \binom{S}{2}$$

Notice that, as we have identified S with the set of its subsets of cardinality 1, if $p(v) = v'$ we can write as well $v' \in p(v)$. We define a relation $\leq_p$ as follows: $v \leq_p v'$ if there exist a chain $v = v_1, ..., v_s = v'$ such that $v_{i+1} \in p(v_i)$. The relation $\leq_p$ is clearly reflexive and transitive.

DEFINITION 2.1. Consider (S, p) be a pair formed by a set and a mapping of the type above. We say that p is a *parental function* if and only if

(1) If $p(v) = \{v', v''\}$, then either $v' \in p(v'')$ or $v'' \in p(v')$ and there is no other vertex v''' such that $p(v''') = \{v', v''\}$.
(2) $\leq_p$ is antisymetric. Hence it defines a partial order in S, that will be called *parental ordering* of (S, p).
(3) The element w in which p is not defined is an absolute maximum for the parental ordering.

If p is a parental function and S is finite we will say that (S,p) is a *parental system*. If p is a parental function and S is infinite we will say that (S,p) is an *infinite parental system*. The element in which p is not defined is called *root*. We refer to the elements of S as the vertices of (S,p)

EXAMPLE 2.2. Let $\pi : X \to ... \to \mathbb{P}^2$ is a blowing up process starting at the line at infinity L. Define S to be the set of irreducible components of π^*L, define p by the property that $E \in p(E')$ if and only if the center of the blowing up giving rise to E' belongs to E. Then (S,p) is a parental system, that will be called *parental system associated with* π.

EXAMPLE 2.3. Let $g(x,y) = y - x^4$. The completion $\overline{C}$ of the curve $C := \{g = 0\}$ to $\mathbb{P}^2$ has only one branch at infinity that meets it at $x_1 = (0:1:0)$. The local equation of $\overline{C}$ at x_1 is $z^3 - x^4 = 0$. Consider the infinite sequence of blowing ups obtained blowing up, at the i-th step, the unique intersection point x_i of the strict transform of $\overline{C}$ with the exceptional divisor E_{i-1} of the previous step. Consider the blowing up process $\pi : X_8 \to ... \to \mathbb{P}^2$ given by the first eight blowing ups of the previous sequence. Then we get a parental system (S,p) with $S = \{E_0 := L, E_1, ..., E_8\}$ and p described by: if $5 \leq i \leq 8$ then $p(E_i) = \{E_{i-1}\}$; if $2 \leq i \leq 4$ then $p(E_i) = \{E_1, E_{i-1}\}$; $p(E_1) = \{E_0\}$.

REMARK 2.4. *Deletion of vertices.* Let (S,p) be a parental system such that S has at least 2 elements, let w be its root and $v \in S$ a minimal element for the ordering $\leq_p$. Define $S' := S \setminus \{v\}$. Then $p(S' \setminus \{w\}) \subset S' \coprod \binom{S'}{2}$ and $(S', p_{|S'})$ is a parental system. It is called the parental system obtained from (S,p) by deletion of v. Note that (S,p) is determined by $(S', p'_{|S'})$ and the value of p at v.

DEFINITION 2.5. Let (S,p) and (S',p') be parental systems. We say that (S',p') is a parental subsystem of (S,p) if and only if $S' \subset S$ and $p_{|S'} = p'$.

Here is an important type of parental subsystems:

REMARK 2.6. Let (S,p) be a parental system and v an element of p. We will denote by $S_{\geq v}$ the set of vertices greater or equal than v by $\leq_p$. Then $(S_{\geq v}, p_{|S_{\geq v}})$ is a parental subsystem of (S,p) and its set of vertices is linearly ordered by $\leq_p$.

DEFINITION 2.7. Let (S,p) be a parental system. Let $w \in S$ be the root. We say that $v \in S$ is of the *i-th generation* if the length of the maximal sequence of the form $v = v_0 \leq_p ... \leq_p v_k = w$ without repeated vertices is i. Define (S_k, p_k) to be the parental system obtained from successively deleting all the vertices of generation strictly bigger than k.

DEFINITION 2.8. A *graph* is a triple (S,E,i) where S and E are finite sets and $i : E \to \binom{S}{2}$, an injective mapping. The sets S and E are called the set of *vertices*, and the set of *edges* respectively; the mapping i is called *incidence mapping*. A *rooted graph* is a graph together with a distinguished vertex w, called the *root*. A *weighted graph* is a graph together with a mapping $\alpha : S \to \mathbb{Z}$ called the *weighting map*.

The data contained in a (rooted) (weighted) graph is best expressed in a picture: draw a white node for the root and a black node for each vertex different from the root. Join with an edge 2 nodes if the vertices that they represent form an element of the image of i. Write down near to each node the value by α of the vertex that it represents.

Given a parental system (S, p) we will associate to it a rooted weighted graph whose set of vertices is S and whose root is the root of (S, p). Suppose that in (S, p) we find vertices of generation at most r. Let w be the root of (S, p). Consider the parental systems $(S_1, p_1), ..., (S_r, p_r) = (S, p)$. We will associate a rooted weighted graph $\mathcal{G}_i$ to each (S_i, p_i) satisfying the following property:

> (*) If $v \in p(v')$ and there is no vertex v'' in S_i such that $p(v'') = \{v, v'\}$ then the vertices v and v' are linked by an edge.

CONSTRUCTION 2.9. We construct the graphs $\mathcal{G}_i$ by induction on i:
Initial step: For $i = 1$ notice that $S_1 = \{w\}$. We associate the graph that consists only of a white node with weight 1. Property $(*)$ is trivial.
Inductive step: Assume that we have associated a rooted weighted graph $\mathcal{G}_i = (w, S_i, E_i, \alpha_i)$ to (S_i, p_i). We associate a rooted weighted graph $\mathcal{G}_{i+1}$ to (S_{i+1}, p_{i+1}) as folows: given a vertex v of the $i + 1$-th generation

(1) If $p(v) = v'$ then we add a black node representing v to $\mathcal{G}_i$, and we join it by an edge to the node representing v'. We define the weights as follows: $\alpha_{i+1}(v) := -1$, $\alpha_{i+1}(v') := \alpha_i(v') - 1$, and, if v'' is a vertex different from v, v', then $\alpha_{i+1}(v'') := \alpha_i(v'')$. The fact that property (*) is preserved is obvious.
(2) If $p(v) = \{v', v''\}$ then we know that either $v' \in p(v'')$ or $v'' \in p(v')$. As v is the unique vertex of S such that $p(v) = \{v', v''\}$ we deduce that the vertices v' and v'' are linked by an edge of $\mathcal{G}_i$. We delete this edge, add a black node representing v and join it with edges to v' and v''. We define the weights as follows: $\alpha_{i+1}(v) := -1$, $\alpha_{i+1}(v') := \alpha_i(v') - 1$, $\alpha_{i+1}(v'') := \alpha_i(v'') - 1$, and if v''' is a vertex different from v, v', v'' then $\alpha_{i+1}(v''') := \alpha_i(v''')$. The fact that property (*) is preserved is obvious.

We do this for each vertex of the $i + 1 - th$ generation and declare that the graph obtained by this procedure is $\mathcal{G}_{i+1}$.

DEFINITION 2.10. Given (S, p) a parental system, we denote by $\mathcal{G}(S, p)$ the graph constructed in the procedure above, and we call it *parental graph* associated to (S, p).

EXAMPLE 2.11. The graph of the parental system considered in Example 2.3 is

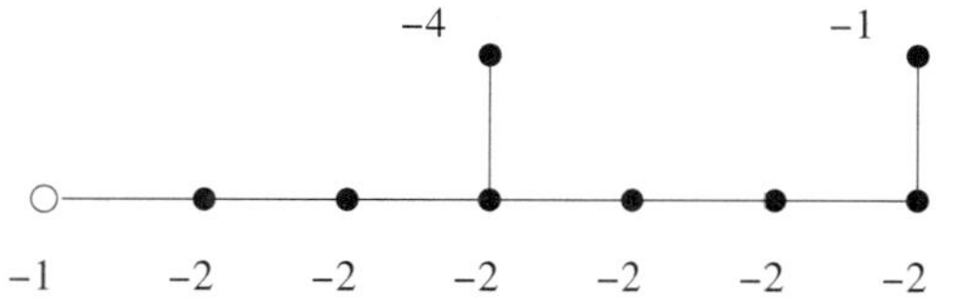

Note that any graph of the parental system associated to a blowing up process is the same that the graph of the blowing up process.

In the following example we will introduce some useful terminology

EXAMPLE 2.12. (*Automorphism graphs.*) For any n, the graph $\mathcal{A}_n$ (see the figure in page 3) is a parental graph whose parental function is $p(v_1) = v_0$, $p(v_i) = \{v_1, v_{i-1}\}$ for $2 \leq i \leq n$ and $p(v_i) = v_{i-1}$ for $n + 1 \leq i \leq 2n - 1$.

For any $r, n_1, ..., n_r$ the graph $\mathcal{A}_{n_1,...,n_r}$ is also a parental graph. It associated parental system (S, p) can be constructed as follows: for any $i \leq r$ let (Σ_i, p_i) be the parental system of the graph $\mathcal{A}_{n_i}$ and denote by $\{v_{i,0}, ..., v_{i,2n_i-1}\}$ the elements of Σ_i. We define the S as

$$S := \cup_{i=1}^{r} \Sigma_i / \sim$$

where $\sim$ identifies $v_{i,2n_i-1}$ with $v_{i+1,0}$ for any $i \leq r-1$; define the function p by $p(v_{i,j}) := p_i(v_{i,j})$ if $j > 0$.

Convention: As all the graphs that appear in this paper will be parental, from now on we will feel free to omit the word "parental" when we speak about parental graphs.

The following easy proposition shows how to pass from graphs to parental systems.

PROPOSITION 2.13. *$\mathcal{G}(S, p)$ determines (S, p). In consequence, if $\mathcal{G}$ is a parental graph then the parental ordering is a canonically defined partial ordering in its set of vertices.*

PROOF. We work by induction on the number of vertices. Inductive step: the set of vertices of a parental system coincides with the set of vertices of its associated graph, hence we only have to describe how to recover the parental function. From the construction of the graph associated with a parental system it is obvious that the vertices that are minimal in the parental ordering are the vertices with weight -1. Also, if an edge connects 2 vertices v, v', then it follows that either $v \in p(v')$ or $v' \in p(v)$. These facts imply that if v is a vertex with weight -1 then $p(v)$ consists of the vertices connected to v by and edge. A parental system is determined by the parental subsystem obtained by the deletion of a minimal vertex v and the value of the parental function at v; therefore if we describe at the level of graphs the operation of deletion of a vertex, then we have reduced the recovery of the parental system of the graph $\mathcal{G}$ to the recovery of the parental system of a graph with less vertices. The proof will be complete after the following remark that provides with the needed description. □

REMARK 2.14. (*Deletion of vertices for graphs*) Given a rooted weighted graph $\mathcal{G}$ associated with a parental system and a vertex v with weight -1, then an easy verification shows that the graph described below is the graph of the parental system obtained by deletion of the vertex v:

(1) If v is attached to a single vertex v' we obtain a new graph by deleting both v and the edge that attaches it to v',and increasing the weight of v' by one.
(2) If v is attached to 2 vertices v'_1, v'_2 then define a new graph by deleting v and the edges that meet it, connecting v'_1 and v'_2 by an edge, and increasing by one the weights of v'_1 and v'_2.

We will need to define graphs decorated with weighted arrows. The motivation for it is the following: if $\mathcal{G}$ is the graph of a blowing up process starting at $L \subset \mathbb{P}^2$ and C is a projective curve, we will be express some properties of the behaviour of C respect to π (such as the intersection multiplicity of the strict transform with the components of $\pi^* L$) by adding weighted arrows to the graph. This will be made precise in the next section.

DEFINITION 2.15. Let $\mathcal{G}$ be a parental graph whose set of vertices is S and whose set of edges is E. A *decoration* on $\mathcal{G}$ consists of four finite sets $\Lambda_1, \Lambda_2, \Lambda_3, \Lambda_4$ of arrows together with:

(1) Four "attaching mappings"

$$\gamma_1 : \Lambda_1 \to S \qquad \gamma_2 : \Lambda_2 \to S$$
$$\gamma_3 : \Lambda_3 \to E \qquad \gamma_4 : \Lambda_4 \to E$$

where γ_1 and $\gamma_3 \coprod \gamma_4 : \Lambda_3 \coprod \Lambda_4 \to E$ are injective and $\gamma_1(\Lambda_1) \bigcap \gamma_2(\Lambda_2)$ is empty.

(2) Four "weighting mappings" mappings

$$\omega_1 : \Lambda_1 \to \mathbb{N} \qquad \omega_2 : \Lambda_2 \to \mathbb{N}^3$$
$$\omega_3 : \Lambda_3 \to \mathbb{N}^2 \qquad \omega_4 : \Lambda_4 \to \mathbb{N}^5$$

A *decorated graph* is a parental graph together with a decoration. If $a \in \Lambda_i$ then $\omega_i(a)$ is called the *weight of* a. Consider an arrow a, the components of its weight receive the following names:

- If $a \in \Lambda_1$ then the only component of $\omega_1(a)$ is said to be *primary.*
- If $a \in \Lambda_2$, the first component of $\omega_2(a)$ is *primary*, the other components are *secondary.* The second component is called the *multiplicity component.*
- If $a \in \Lambda_3$ then the only two components of $\omega_3(a)$ are said to be *primary.*
- If $a \in \Lambda_4$, the first two components of $\omega_4(a)$ are *primary*, the other components are *secondary.* The third component is called the *multiplicity component.*

NOTATION 2.16. Let

$$\Lambda := \coprod_{i=1}^{4} \Lambda_i$$

and define

$$\gamma := \Lambda \to S \bigcup E \quad \omega := \Lambda \to \mathbb{N} \coprod \mathbb{N}^3 \coprod \mathbb{N}^2 \coprod \mathbb{N}^5$$

such that their respective restrictions to Λ_i, for $i \leq 4$, are γ_i and ω_i. We will say that γ is the attaching mapping of the decoration and that ω is the weighting map of the decoration. From now on pair $(\mathcal{G}, \Lambda)$ will denote a graph $\mathcal{G}$ together with a decoration.

The information contained in a decorated graph $(\mathcal{G}, \Lambda)$ is expressed graphically as follows: draw the graphical representation of $\mathcal{G}$, for $1 \leq i \leq 4$ and for each element $a \in \Lambda_i$ attach an arrow to the element of the graph $\mathcal{G}$ designated by $\gamma_i(a)$, and write down the weight $\omega_i(a)$ near the arrow corresponding to a.

EXAMPLE 2.17. An example of decorated graph is

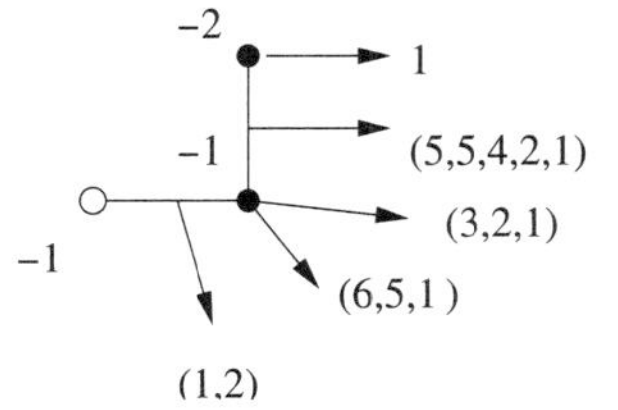

The arrow with weight 1 belongs to Λ_1; the arrow with weight $(1, 2)$ belongs to Λ_3; the arrows with weights $(6, 5, 1)$ and $(3, 2, 1)$ belong to Λ_2; the arrow with weight $(5, 5, 4, 2, 1)$ belongs to Λ_4.

Construction 2.18. (Deletion of the smallest vertex). Consider a decorated graph $(\mathcal{G}, \Lambda)$, let S and E be its sets of vertices and edges, let v be a vertex with weight -1. Denote by $\mathcal{G}'$ the graph obtained from $\mathcal{G}$ by deletion of the vertex v, and by S' and E' its sets of vertices and edges. Then the following procedure describe a method for constructing a decoration Λ' on $\mathcal{G}'$ starting from Λ:

- If $e \in E$ is an edge of $\mathcal{G}$ not meeting v, then it is also an edge $e \in E'$. If there is an arrow $a \in \Lambda$ with weight $\omega_i(a)$ attached to $e \in E$, attach an arrow $a \in \Lambda'$ to $e \in E'$ with the same weight.
- If $v' \in S$ is a vertex not connected to v, then it is also a vertex $v \in S'$. For each arrow $a \in \Lambda$ attached to $v' \in S$ attach an arrow $a \in \Lambda'$ to $v' \in S'$ with the same weight.
- If v is connected by an edge e to a single vertex v', then after the deletion of the vertex v the edge e disappears and v' becomes a vertex of S'. We attach the following arrows to v': for each arrow of Λ attached to $v' \in S$ we attach an arrow of Λ' with the same weight to $v' \in S'$. In addition, if there are arrows of Λ attached to e or v, we attach one more arrow to v' with weight (s, s', s''), where (s, s', s'') is constructed as follows: let $a_0 \in \Lambda$ be the arrow attached to e (if it exists) and let t_0, t'_0 be the first two coordinates of its weight; if a_0 does not exist define $t_0 = t'_0 = 0$. Let $a_1, ..., a_k$ be the arrows attached to v, and let $t_1, ..., t_k$ be the first coordinates of their weights, then

$$s := t'_0 + \sum_{i=0}^{k} t_i \quad s' := t'_0 + \sum_{i=1}^{k} t_i \quad s'' := t'_0$$

- If v is connected to two vertices v' and v'', by edges e' and e'' respectively and $v' \geq_p v''$, then, after the deletion of the vertex v, the edges e' and e'' disappear and the vertices v', v'' become connected by an edge $e \in E'$. We add the following arrows: for each arrow of Λ attached to $v' \in S$ we attach an arrow of Λ' to $v' \in S$ with the same weight; for each arrow of Λ attached to $v'' \in S$ we attach an arrow of Λ' to $v'' \in S'$ with the same weight; if there are arrows of Λ attached to e' ,e'' or v, we attach one more arrow of Λ' to e whose weight $(s_1, s_2, s_3, s_4, s_5)$ is constructed as follows: let $a'_0 \in \Lambda$ be the arrow attached to e' and let t_0, t'_0 be the first two coordinates of its weight, let $a''_0 \in \Lambda$ be the arrow attached to e'' and let u_0, u'_0 be the first two coordinates of its weight; if a'_0 (resp. a''_0) does not occur then define $t_0 = t'_0 = 0$ (resp $u_0 = u'_0 = 0$). Let $a_1, ..., a_k$ be the arrows attached to v, and let $t_1, ..., t_k$ be the first coordinates of their weights, then

$$s_1 := t_0 + t'_0 + u'_0 + \sum_{i=1}^{k} t_i \quad s_2 := t'_0 + u_0 + u'_0 + \sum_{i=1}^{k} t_i$$

$$s_3 := t'_0 + u'_0 + \sum_{i=1}^{k} t_i \quad s_4 := t'_0 \qquad s_5 := u'_0$$

Example 2.19. We will succesively delete the vertices of the graph of Example 2.17

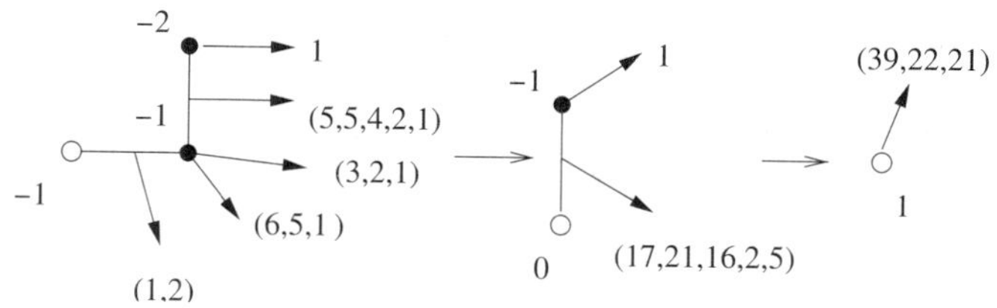

The following definitions introduces the two main classes of graphs of this paper:

DEFINITION 2.20. Let $\mathcal{G}$ be a parental graph. We say that it is $\mathcal{A}$-constructible if and only if given any vertex of $v \in \mathcal{G}$ then there exist $r, n_1, ..., n_r$ such that the parental system $(S_{>v}, p_{S_{>v}})$ can be embedded in the parental system of the graph $\mathcal{A}_{n_1,...n_r}$ as a parental subsystem.

NOTATION 2.21. Consider the graph $\mathcal{A}_{n_1,...,n_r}$. Clearly it is obtained taking the union of the graphs $\mathcal{A}_{n_1}$,..., $\mathcal{A}_{n_r}$ and, for any $i < r$, identifying the smallest vertex in the parental ordering of $\mathcal{A}_i$ with the root of $\mathcal{A}_{i+1}$. We will give the use the following terminology when refering to $\mathcal{A}_{n_1,...,n_r}$:

- If a vertex v is in $\mathcal{A}_i$ we say that v is in the i-th floor (note that the root of the i-th floor is also the smallest vertex of the $i-1$-th floor and viceversa.
- We say that the i-th floor is a floor of length n_i.
- Let $v_{i,0}, v_{i,1}, ..., v_{i,2n_i-1}$ be the set of vertices of the i-th floor ordered decreasingly respect to $\leq_p$. We will say that: $v_{i,0}$ is the root of the i-th floor, $v_{i,2n_i-1}$ is the top of the i-th floor, $v_{i,1}$ is the free central vertex, $v_{i,2}, ..., v_{i,n_r-1}$ are the fixed lateral vertices, v_{i,n_r} is the fixed central vertex and $v_{i,n_i+1}, ..., v_{i,2n_r-1}$ are the free lateral vertices. The fixed vertices are the fixed lateral vertices together with the fixed central vertex.

NOTATION 2.22. Let $\mathcal{G}$ be a $\mathcal{A}$-constructible graph and let v a vertex of it. Let $r, n_1, ..., n_r$ be the smallest numbers such that $(S_{>p}, p_{S_{>p}})$ can be embedded in the parental system of the graph $\mathcal{A}_{n_1,...n_r}$. Then we will say that v is a vertex of the r-th floor of $\mathcal{G}$. We say that v is the top, the free central vertex, a fixed vertex or a free lateral vertex if and only if it is so as a vertex of $\mathcal{A}_{n_1,...,n_r}$.

DEFINITION 2.23. Let $(\mathcal{G}, \Lambda)$ be a decorated graph. Let (S, p) be its associated parental system. We say that it is a *behaviour graph* if and only if

(1) The graph $\mathcal{G}$ is $\mathcal{A}$-constructible.
(2) For any vertex v there exists another vertex v' smaller or equal than v by $\leq_p$ such that there is an arrow attached either to v' or to and edge that meets v'.
(3) Both Λ_2 and Λ_4 are empty.
(4) The arrows of Λ_1 only can be attached to free central vertices.
(5) The arrows of Λ_3 cannot be attached to edges that meet a free central vertex.

We say that $(\mathcal{G}, \Lambda)$ is a *partial resolution graph* if and only if it can be obtained by deletion of vertices from a behaviour graph.

DEFINITION 2.24. Let $(\mathcal{G}, \Lambda)$ be a partial resolution graph and v a vertex different from the root. We define the *total multiplicity* of v and we denote it by $\mathrm{mtot}(v)$ in the following inductive way:

- If v is of the first generation consider the graph $\mathcal{G}_0$ obtained by deleting all the vertices except the root. There is an arrow a in $\mathcal{G}_0$ that corresponds to the deletion of v. The total multiplicity of v is defined to be equal to the multiplicity component of the decoration of a.
- Assume that the total multiplicity is defined for any vertex of generation smaller than i. Let v be a vertex of the i-th generation. Let $\mathcal{G}_{i-1}$ be the graph obtained from the deletion of all the vertices of generation i or greater. There is an arrow a in $\mathcal{G}_{i-1}$ that corresponds to the deletion of v, let t be the value of the multiplicity component of the weight of a. We define the total multiplicity of v by the formula:

$$\mathrm{mtot}(v) := t + \sum_{v' \in p(v)} \mathrm{mtot}(v')$$

We will define the *degree* of the graph $\mathcal{G}$ and denote it by $\deg(\mathcal{G})$ to the sum of all the first components of the weights of the arrows of $\mathcal{G}_0$.

DEFINITION 2.25. Let (S, p) and (S', p') parental systems. An isomorphism from (S, p) to (S', p') is a bijection $\varphi : S \to S'$ that satisfies $p' \circ \varphi = \varphi \circ p$. Let $\mathcal{G} = (w, S, E, i, \alpha)$ and $\mathcal{G}' = (w', S', E', i', \alpha')$ be rooted weighted graphs. An isomorphism $\beta : \mathcal{G} \to \mathcal{G}'$ is a pair (φ, ψ) where $\varphi : S \to S'$ and $\psi : E \to E'$ are bijections such that

$$\varphi(w) = w' \qquad \alpha' \circ \varphi = \alpha \qquad i' \circ \psi = \binom{\varphi}{2} \circ i$$

where $\binom{\varphi}{2} : \binom{S}{2} \to \binom{S}{2}$ takes $\{v, v'\}$ to $\{\varphi(v), \varphi(v')\}$. In other words β is is a bijection from the set of vertices of $\mathcal{G}$ to the set of vertices of $\mathcal{G}'$ that sends the root to the root, vertices to vertices with the same weight, and vertices connected by an edge to vertices connected by an edge.

Notice that both notions of isomorphism coincide via the identification between parental systems and parental graphs that we have made.

DEFINITION 2.26. Let $(\mathcal{G}, \Lambda)$, $(\mathcal{G}', \Lambda')$ be decorated graphs. An isomorphism $\xi : (\mathcal{G}, \Lambda) \to (\mathcal{G}', \Lambda')$ is a pair (ξ_1, ξ_2) where $\xi_1 = (\varphi, \psi) : \mathcal{G} \to \mathcal{G}'$ is an isomorphism and ξ_2 is a bijection from the set of arrows Λ to the set of arrows Λ' such that it takes each arrow $a \in \Lambda$ to an arrow $a' \in \Lambda'$ with the same weight and attached to the transformation by ξ_1 of the attaching point of a.

DEFINITION 2.27. Let $(\mathcal{G}, \gamma)$ be a decorated graph, x and y two vertices, edges, or arrows, we say that x and y are combinatorially equivalent, and we denote it by $x \sim_c y$, if there is an automorphism ξ of $(\mathcal{G}, \Lambda)$ taking x to y. Obviously this is an equivalence relation.

REMARK 2.28. If e, e' are combinatorially equivalent edges and e (resp. e') connects the vertices v_1, v_2 (resp. v_1', v_2') with $v_1 \leq_p v_2$ (resp. $v_1' \leq_p v_2'$) then $v_1 \sim_c v_1'$ and $v_2 \sim_c v_2'$. Let a_1, a_2 be two combinatorially equivalent arrows, then $\gamma(a_1) \sim_c \gamma(a_2)$.

2.2. The partition

Let X be a smooth algebraic surface. We say that $\pi : Y \to X$ is a modification of X if and only if it is a proper and birational morphism with Y smooth (in particular, an isomorphism is a modification). It is classical that any proper modification between two smooth surfaces is a composition of blowing ups at points. We consider the set M_X of all the proper modifications of $Y \to X$ with Y smooth. Consider the set

$$\mathcal{X}'_X := \coprod_{\pi:Y\to X\in M_X} Y$$

Let $\phi : Y \to Y'$ be a modification. Define U_ϕ to be the maximal open subset of Y' such that the restriction $\phi : \phi^{-1}(U_\phi) \to U_\phi$ is an isomorphism. Define the following relation on $\mathcal{X}'_X$: let $y, y' \in \mathcal{X}'_X$, we know that $y \in Y$, $y' \in Y'$ for two modifications $\pi : Y \to X$ and $\pi' : Y' :\to X$; we say that y is equivalent to y' and we denote it by $y \sim y'$ if and only if there exist modifications $\phi : Y'' \to Y$ and $\phi' : Y'' \to Y'$ and such that $y \in U_\phi$, $y' \in U_{\phi'}$ and there is $y'' \in Y''$ satisfying $\phi(y'') = y$ and $\phi'(y'') = y'$. It is easy to show that " $\sim$" is an equivalence relation. Define the set $\mathcal{X}_X$ of infinitely near points to X as

$$\mathcal{X}_X := \mathcal{X}'_X / \sim$$

Let $y \in Y$, $y' \in Y'$ where $\pi : Y \to X$, $\pi' : Y' \to X$ are modifications. If $y \sim y'$ it is clear that $\pi(y) = \pi'(y')$.

DEFINITION 2.29. Let L be an algebraic subset of X and $[x] \in X$. We say that $[x]$ is infinitely near to L if $\pi(y) \in L$ for any representative $y \in Y \xrightarrow{\pi} X$ of $[x]$. Let $\dot{\mathcal{X}}_L$ be the subset of $\mathcal{X}$ formed by points infinitely near to L; define $\mathcal{X}_L$ as the disjoint union $\dot{\mathcal{X}}_L \coprod \{w_L\}$, where $\{w_L\}$ is a set of one element which is called the root.

DEFINITION 2.30. Let $\pi : X_n \xrightarrow{\pi_n} \ldots \xrightarrow{\pi_1} X_0 = X$ be a blowing up process. Let $x_i \in X_{i-1}$ be the center of π_i. We say that $I_\pi := \{[x_1], \ldots, [x_n]\}$ is the *set of infinitely near points associated with the blowing up process.*

Let $L \subset X$ be an algebraic subset. Let $I \subset \mathcal{X}_L$. We say that I is a *saturated* set of infinitely near points to L if it contains ω_L and $I \setminus \{\omega_L\}$ is the set of infinitely near points associated with a blowing up process whose first center is in L. We will write for short "I is a s.s.i.n.p. to L".

If I is a s.s.i.n.p. to L and, for a blowing up process π, it is equal to $\{\omega_L\} \cup I_\pi$, then we say that *π is a model for I.*

CONSTRUCTION 2.31. Consider $L \subset X$, an algebraic subset. We define a mapping

$$p : \mathcal{X}_L \setminus \{w_L\} \to \mathcal{X}_L \coprod \binom{\mathcal{X}_L}{2}$$

such that (X_L, p) is an infinite parental system: let $[x] \in \mathcal{X}_L$, there is a minimal blowing up process

$$X_n \xrightarrow{\pi_n} \ldots \xrightarrow{\pi_1} X_0 = X$$

such that, if x_i denotes the center of π_i then $[x] = [x_n]$. Define

- $p[x] = \{[x_i], [x_j]\}$ if and only if $\{x_n\} = E_i \bigcap E_j$.
- $p[x] = \{[x_i]\}$ if and only if $x_n \in \dot{E}_i$

It is easy to check that p is a parental function.

REMARK 2.32. Let $J \subset \mathcal{X}_L$ be finite. Then J is a s.s.i.n.p. to L if and only if the image of $J \setminus \{w_L\}$ by p is contained in $J \coprod \binom{J}{2}$. Hence if J is a s.s.i.n.p. to L and p_J is the restriction of p to J then (J, p_J) is a parental system which we call the *parental system associated with* J. Define the *graph associated with* J to be the graph $\mathcal{G}_J := \mathcal{G}(J, p_J)$ associated with the parental system (J, p_J).

REMARK 2.33. Let J be a s.s.i.n.p. to L, and consider a blowing up process π that is a model for J. Then $\mathcal{G}_J$ is isomorphic to the graph of π.

From now on we will be concerned with the case in which $X = \mathbb{P}^2$ and L is the line at infinity.

NOTATION 2.34. Consider $f \in \mathbb{C}[x, y]$; its zero locus is an affine plane curve which will be denoted by C. Embed $\mathbb{C}^2$ into $\mathbb{P}^2$ by considering a projective coordinate system (X, Y, Z) and declaring that $(X/Z, Y/Z)$ are the coordinates of $\mathbb{C}^2$, let $F \in \mathbb{C}[X, Y, Z]$ be the homogeneization of f with respect to the variable Z, the projective curve $\overline{C}$ defined by F is the projective completion of the affine curve C. By abuse of language, when it does not lead to confusion, denote both C and $\overline{C}$ by C and both f and F by f. Consider a blowing up process

$$\pi : X_n \xrightarrow{\pi_n} \dots \xrightarrow{\pi_1} \mathbb{P}^2$$

The strict transform of $\overline{C}$ by the composition $\pi_1 \circ \dots \pi_i$ will be denoted by $\tilde{C}_i$.

CONSTRUCTION 2.35. Given $f \in \mathbb{C}[x, y]$ we define $\mathcal{J}_f$ to be the following set of infinitely near points to L: consider $[x] \in \mathcal{X}_L$, there is a minimal blowing up process

$$X_n \xrightarrow{\pi_n} \dots \xrightarrow{\pi_1} X_0 = \mathbb{P}^2$$

such that, if x_i denotes the center of π_i then $[x] = [x_n]$; clearly there is a unique representative x of $[x]$ which is a point of X_{n-1}; then $[x] \in \mathcal{J}_f$ if and only if the strict transform of C by $\pi_1 \circ \dots \circ \pi_{n-1}$ meets x.

Our aim is to construct the maximal saturated subset of infinitely near points of $\mathcal{J}_f$ such that each of the points of points is also an infinitely near point of a blowing up associated with an automorphism of $\mathbb{C}^2$. We do it as follows: let

$$\mathcal{I} := \bigcup_{k \in \mathbb{N}} \Big(\bigcup_{(n_1, \dots, n_k) \in \mathbb{N}^k} \big(\cup_{\pi \in A_{n_1, \dots, n_k}} I_\pi \big) \Big) \tag{2.1}$$

Finally we define

$$I_f := (\mathcal{J}_f \cap \mathcal{I}) \cup \{\omega_L\} \tag{2.2}$$

It is easy to verify that the image of $I_f \setminus \{w_L\}$ by p is contained in $I_f \coprod \binom{I_f}{2}$, hence to show that I_f is a s.s.i.n.p. to L we only have to verify that it is finite. □

LEMMA 2.36. *For any* f *the set* I_f *is finite.*

PROOF. After a finite number of blowing ups centered in points of $\mathcal{J}_f$ any singularity of C at infinity achieves its embedded resolution. Let this number be s. If $J \subset \mathcal{J}_f$ is saturated and $v \in J$ a vertex of the r-th generation with $r > s$, then the set $p(v)$ has only one vertex. If we impose also that J is a subset of $\mathcal{I}$ linearly ordered by $\leq_p$, because of the shape of the parental systems associated with automorphisms, we deduce that $\sharp J \leq 2s$. We have shown that any chain of I_f that is linearly ordered by the parental ordering has at most $2s$ vertices. As the number of maximal linearly ordered chains of $\mathcal{J}_f$ is bounded by the number of branches of C at infinity the lemma is proved. □

EXAMPLE 2.37. Let $f(x,y) = x^5 - y^3$. The curve defined by its zero fibre has only one branch at infinity, defined locally by the equation $x^5 - z^2$. Hence the infinitely near points that are centers of blowing ups of the embedded resolution of the singularity at infinity are totally ordered by the parental ordering. The set I_f consists of the root w_L together with the two first blowing up centers of the embedded resolution of the singulatity at infinity. The set I_f has three points.

EXAMPLE 2.38. Let $g(x,y) = y - x^4$. The curve defined by its zero fibre has, as in the case before, only one branch at infinity. The set I_g consists of the root w_L together with all the centers of the blowing up process obtained by adding to the embedded resolution of the singularity at infinity three successive blowing ups, centered at the intersection of the strict transform with the exceptional divisor. The set I_g has nine points.

DEFINITION 2.39. Let $f \in \mathbb{C}[x,y]$. The set I_f will be called *behaviour set at infinity* of f, or for short *behaviour set* of f. A model of I_f will be called *behaviour resolution* of f. A *partial resolution set* I for f is a s.s.i.n.p. to L such that $I \subset I_f$. A *partial resolution* of f is a model of a partial resolution set.

Notice that the term partial resolution gets a meaning different from the meaning that it has usually in the literature. Next we introduce a definition whose terminology will become more motivated later in this paper.

DEFINITION 2.40. Let I_f be the behaviour set associated with a polynomial $f \in \mathbb{C}[x,y]$. Let (I_f, p) be its associated parental system. We say that $x \in I$ is free if $\sharp p(x) = 1$, and that x is fixed if $\sharp p(x) = 2$. We say that $x \in S$ *supports moduli* (for short x is *s.m.*) if there exists $x' \in I_f$ such that $p(x') = x$, otherwise we say that x *does not support moduli* (for short x is *n.s.m.*).

If $I \subset I_f$ is such that I is the set of infinitely near points of a partial resolution then, by definition, $x \in I$ is said to be *free ,fixed, s.m., n.s.m.* if it is so as an element of I_f.

The fixed points are also called *satellite* points in the literature.

EXAMPLE 2.41. Suppose that f and g are like in Examples 2.37 and 2.38, let $\{\omega, x_1, x_2\}$ and $\{\omega', y_1, ..., y_8\}$ be the points of I_f and I_g ordered by the parental order. Then ω_L, ω', y_4, y_5, y_6, y_7 are *s.m*, the rest of the points are *n.s.m*; the free points are x_1, y_1, y_5, y_6, y_7, y_8; the fixed points are x_2, y_2, y_3, y_4.

REMARK 2.42. Let f, I_f and I as in the definition above. The graph $\mathcal{G}_I$ is $\mathcal{A}$-constructible, and hence, the terminology of notation 2.22 is valid for it, and is consistent with the definition above. Note that a free central vertex cannot be *s.m.*

NOTATION 2.43. Let C, C' two divisors of an smooth surface. We will denote by $I_p(C, C')$ the intersection multiplicity of C and C' at p.

CONSTRUCTION 2.44. Consider $f \in \mathbb{C}[x,y]$ and let $I := \{x_1, ..., x_n\} \subset I_f$ be a partial resolution set. Choose π, a model for I. Let $\tilde{C}$ the strict transform of C by π. We add a decoration to $\mathcal{G}_I$ as follows:

- If $\tilde{C}$ meets the intersection $q := E_i \bigcap E_j$ for $i, j \in \{0, ..., n\}$, with $x_j \leq_p x_i$ and $q \in I_f$. Then we attach an arrow to the edge connecting the vertices associated with x_i and x_j and we label it with the quintuple

$$(I_q(E_i, \tilde{C}), I_q(E_j, \tilde{C}), \mathrm{mult}_q(\tilde{C}), I_{E_i \cap E}(E, \tilde{\tilde{C}}), I_{E_j \cap E}(E, \tilde{\tilde{C}}))$$

where E and $\tilde{\tilde{C}}$ are the exceptional divisor and the strict transform of $\tilde{C}$ respect to the blowing up centered at q.

- If $\tilde{C}$ meets the intersection $q := E_i \bigcap E_j$ for $i, j \in \{0, ..., n\}$ with $x_j \leq_p x_i$ and q is not in I_f, then attach an arrow to the edge connecting the vertices associated with x_i and x_j and label it with the pair
$$(I_q(E_i, \tilde{C}), I_q(E_j, \tilde{C}))$$
- If $\tilde{C}$ meets $\dot{E}_i$ and x_i is *supporting moduli* then for each point $q \in \tilde{C} \bigcap \dot{E}_i$ we attach an arrow to the vertex representing x_i and label it with the triple
$$(I_q(E_i, \tilde{C}), \mathrm{mult}_q(\tilde{C}), I_{E_i \cap E}(E, \tilde{\tilde{C}}))$$
where E and $\tilde{\tilde{C}}$ are like in the first case.
- If $\tilde{C}$ meets $\dot{E}_i$ and x_i is *non supporting moduli* then we attach a *single* arrow (even when $\tilde{C}$ meets $\dot{E}_i$ in more than one point) to the vertex representing x_i and we label it with the number
$$\sum_{q \in \dot{E}_i \cap \tilde{C}} I_q(E_i, \tilde{C})$$

It is easy to prove that the procedure above does not depend on the model of I.

DEFINITION 2.45. Let $f \in \mathbb{C}[x, y]$ and I a partial resolution set. We define the *graph associated with the pair* (f, I) to be the decorated graph $(\mathcal{G}_I, \Lambda)$ where $\mathcal{G}_I$ is the graph associated with I and Λ is the decoration defined in the last construction above. We say that $(\mathcal{G}, \Lambda)$ is a *partial resolution graph for* f if and only if it is the graph associated with a pair (f, I). We define the *behaviour graph* of f to be the graph $(\mathcal{B}, \Lambda)$ associated with the pair (f, I_f).

REMARK 2.46. Let $f \in \mathbb{C}[X, Y]$, and I,I' partial resolution sets of f such that $I' \subset I$. Let $(\mathcal{G}, \Lambda)$, $(\mathcal{G}', \Lambda')$ be the graphs of I and I' respectively. Then the graph $(\mathcal{G}', \Lambda')$ can be obtained from $(\mathcal{G}, \Lambda)$ by deletion of the vertices corresponding to the points of $I \setminus I'$.

NOTATION 2.47. We will usually denote behaviour graphs by the symbol $\mathcal{B}$ and variants of it in which we add subindices or superindices. The symbol $\mathcal{B}_f$ will denote the behaviour graph of f.

EXAMPLE 2.48. The behaviour graphs of the polynomials f and g defined in Examples 2.37 and 2.38 are respectively

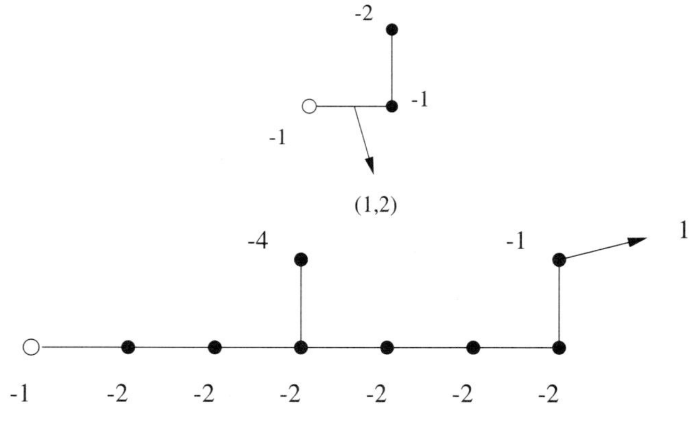

PROPOSITION 2.49. *Let $f \in \mathbb{C}[X,Y]$, and let*

$$\pi : X_n \xrightarrow{\pi_n} \dots \xrightarrow{\pi_1} \mathbb{P}^2$$

be a blowing up process which is a model for I_f. Denote by x_i the center of the π_i and let σ_i be the composition $\pi_1 \circ \dots \circ \pi_i$. Let $\mathcal{B}_f$ be the behaviour graph of f and let v_i the vertex representing x_i.

(1) *The multiplicity of the total transform $\sigma_{i-1}^* C$ at x_i equals* $\operatorname{mtot}(v_i)$.
(2) *The multiplicity of the strict transform $\tilde{C}_{i-1}$ at x_i equals the multiplicity component of the arrow that we obtain if we delete v_i.*

Let S be any subset of the set of vertices of $\mathcal{B}$ such that if v belongs to S then any other v' satisfying $v' \leq_p v$ belongs also to S. Let $\mathcal{G}$ be the graph obtained from $\mathcal{B}_f$ by deletion of the vertices of S. Let π' be a model for the set of infinitely near points corresponding to the vertices of $\mathcal{G}$, and let $\tilde{C}'$ be the strict transform of C by π'. Consider any arrow a of $\mathcal{G}$.

- *If a is attached to an edge connecting two vertices v_i, v_j such that $v_j \leq_p v_i$ then the intersection multiplicities*

$$I_{E_i \cap E_j}(\tilde{C}', E_i) \quad I_{E_i \cap E_j}(\tilde{C}', E_j)$$

are respectively equal to the two primary components of the weight of a. If $a \in \Lambda_4$ then $x := E_i \cap E_j \in I_f$. Let π_x denote the blowing up at x, E_x its exceptional divisor and $\tilde{C}'$ the strict transform of $\tilde{C}'$ by π_x. Then the intersection multiplicities

$$I_{E_i \cap E_x}(\tilde{\tilde{C}}', E_x) \quad I_{E_j \cap E_x}(\tilde{\tilde{C}}', E_x)$$

are respectively equal to the two last secondary components of the weight of a.

- *If a is attached to a s.m. vertex v_i then a has come from the deletion of a vertex v. Let x be the point of I_f associated with it. Let π_x denote the blowing up at x, E_x its exceptional divisor and $\tilde{\tilde{C}}'$ the strict transform of $\tilde{C}'$ by π_x. Then the intersection multiplicities*

$$I_x(\tilde{C}', E_i) \quad I_{E_i \cap E_x}(\tilde{\tilde{C}}', E_x)$$

are respectively equal to the two components of the weight of a that are not the multiplicity component.

- *If a is attached to a n.s.m. vertex v_i then the sum*

$$\sum_{y \in \tilde{C}' \cap \dot{E}_i} I_y(\tilde{C}', E_i)$$

equals the weight of a.

Finally

$$\deg(f) = \deg(\mathcal{B}_f)$$

PROOF. The proof consits in comparing how we construct the weights of the arrows in the procedure of deletion of vertices and the following well known formulae that explain the behaviour of intersection multiplicity under blowing up: let C, C' be two curves meeting at a point x in a smooth surface X, let π be the blowing

up of X at x, let E denote the exceptional divisor, and $\tilde{C}$, $\tilde{C}'$ denote the strict transforms of C and C', then
(2.3)
$$I_x(C,C') = \sum_{y\in\tilde{C}\cap\pi^*C'} I_y(\tilde{C},\pi^*C') = \sum_{y\in\tilde{C}\cap\tilde{C}'} I_y(\tilde{C},\tilde{C}') + \mathrm{mult}_p(C') \sum_{y\in\tilde{C}\cap E} I_y(\tilde{C},E)$$

For the last assertion we use Bezout theorem to ensure that
$$\deg(f) = \sum_{y\in C\cap L} I_y(C,L)$$

□

Finally we are able to define our partition on $\mathbb{C}[x,y]$:

DEFINITION 2.50. Let $\mathcal{B}$ be a behaviour graph. We define $X_{\mathcal{B}}$ to be the set of polynomials whose behaviour graph is isomorphic to $\mathcal{B}$. We define a partition on $\mathbb{C}[x,y]$ imposing that the subsets of the partition are the $X_{\mathcal{B}}$ where $\mathcal{B}$ ranges in the set of behaviour graphs.

Note that a priori some of the subsets of the partition could be empty: we do not know if any behaviour graph is realizable by a polynomial. We will show later that this is the case. Until now we know nothing about the geometry of each subset of the partition $X_{\mathcal{B}}$. We will show that they are smooth algebraic varieties that admit a very pleasant description in terms of the graph $\mathcal{B}$.

2.3. Automorphisms and orderings of graphs

Given a certain behaviour graph, when one tries to study the geometric properties of the subsets $X_{\mathcal{B}}$ (see Definition 2.50) it turns out that it is much more convenient if the graph $\mathcal{B}$ does not have symmetry at all. If $\mathcal{B}$ has symmetry, we will get rid of it introducing a certain type of orderings in the sets of vertices and arrows of $\mathcal{B}$.

DEFINITION 2.51. Let $(\mathcal{G},\Lambda)$ be a decorated graph. Consider the ordering $\leq_p$ of the vertices of the graph and the combinatorial equivalence relation $\sim_c$. An *admissible ordering* $\leq_a$ is a total ordering of each $\sim_c$-equivalence class of vertices or arrows, such that:

(1) If v_1, v_2, v_1', v_2' are vertices satifying $v_1 \sim_c v_2$, $v_1' \sim_c v_2'$, and $v_1 \leq_p v_1'$, $v_2 \leq_p v_2'$, then $v_1 \leq_a v_2 \Rightarrow v_1' \leq_a v_2'$.
(2) If a_1, a_2 are $\sim_c$-equivalent arrows then $a_1 \leq_a a_2 \Rightarrow \gamma(a_1) \leq_a \gamma(a_2)$.

CONSTRUCTION 2.52. Given a decorated graph $(\mathcal{G},\Lambda)$, we construct an admissible ordering $\leq_a$ on it: notice that two combinatorially equivalent vertices are necessarily of the same generation. Hence to define $\leq_a$ in the $\sim_c$-classes of vertices we can proceed inductively on the order of generation of the elements of each $\sim_c$-class: at the initial step the only vertex of the 0-th generation is the root, so there is nothing to order. Assume that we have defined $\leq_a$ for all vertices of generation smaller or equal than k. Consider V, a $\sim_c$ *−class* of vertices of the $k+1$-th generation; impose the following partial ordering in V: let $v_1, v_2 \in V$, then $v_1 < v_2$ if there exist $v_1' \geq_p v_1$ and $v_2' \geq_p v_2$ such that $v_1' <_a v_2'$. Complete $\leq$ to a total ordering in V and declare it to be the restriction of $\leq_a$ to V. Do this for every $\sim_c$ class of vertices of the $k+1$-th generation.

Let A be a $\sim_c$-class of arrows. The admissible ordering defined above in the set of vertices induces a partial ordering in A in a similar manner as in the inductive step of the construction of $\leq_a$ for vertices. We complete this partial ordering to a total ordering in A and declare it to be the restriction of $\leq_a$ to A.

REMARK 2.53. Consider a decorated graph $(\mathcal{G}, \Lambda)$ with an admissible ordering $\leq_a$. Then the only automorphism of $\mathcal{G}$ that preserves $\leq_a$ is the identity.

The above remark means heuristically that giving an admissible ordering to a decorated graph makes all the vertices and arrows combinatorially different from each other. Given an admissible order in a graph, to give a total ordering of the sets of vertices and arrows all we have to do is to give a total ordering in the set of $\sim_c$-classes, and for this we do not need to add more information from the combinatorial point of view, (different $\sim_c$-classes are distinguished from each other). This suggests the existence of a canonical way of ordering the $\sim_c$-classes of any graph. The next proposition states that this a canonical way of ordering graphs exists, and that it has some good properties respect to the parental order and the operation of deletion of the last vertex. As we only need to work with $\mathcal{A}$-constructible graphs, we will restrict our construction to this kind of graphs (although it can be given in more generality). As the techniques of the proof are completely independent of the subsequent developments of the paper, we give it an appendix.

DEFINITION 2.54. Denote by $\mathbf{G}$ the set of decorated $\mathcal{A}$-constructible graphs.

PROPOSITION 2.55. *There exist a canonical order $\leq_c$ in each graph $(\mathcal{G}, \Lambda)$ of $\mathbf{G}$ and an ordering $\leq_g$ on $\mathbf{G}$ such that*

(1) *$\leq_c$ extends the parental ordering.*
(2) *Two vertices or arrows of the graph are $\sim_c$-equivalent if and only if they are not compared by $\leq_c$. Hence $\leq_c$ orders completely the sets of $\sim_c$-classes of vertices and arrows.*
(3) *If we have two graphs that are not compared by $\leq_g$ then we know that they are isomorphic. Hence $\leq_g$ gives a total ordering in the set of graphs modulo isomorphism.*
(4) *Consider $\leq_a$, an admissible ordering. Define a total ordering $\leq_t$ in the sets of vertices and arrows of $(\mathcal{G}, \Lambda)$ by imposing that given x and y , two vertices or arrows, then $x \leq_t y$ if and only if either $x \sim_c y$ and $x \leq_a y$ or $x <_c y$. Let $(\mathcal{G}_0, \Lambda_0)$ the graph obtained from the deletion of the last vertex of $(\mathcal{G}, \Lambda)$ and denote by $\leq_c'$ the canonical order on it. Let $\leq_t'$ be the total order on $(\mathcal{G}_0, \Lambda_0)$ obtained by restriction of $\leq_t$. Then there exists an admissible order $\leq_a'$ in $(\mathcal{G}_0, \Lambda_0)$ such that $\leq_t'$ is the total ordering obtained from $\leq_c'$ and $\leq_a'$.*
(5) *$(\mathcal{G}_0, \Lambda_0) <_g (\mathcal{G}, \Lambda)$.*
(6) *Let $\leq_c$ be the canonical order on $(\mathcal{G}, \Lambda)$ and $\leq_c'$ be the canonical order on $\mathcal{G}$ (the graph without the decoration). If v and v' are vertices of $\mathcal{G}$ such that $v <_c' v'$ then $v <_c v$.*
(7) *Let $(\mathcal{G}, \Lambda)$, $(\mathcal{G}', \Lambda')$ be decorated graphs and $\mathcal{G}$, $\mathcal{G}'$ the same graphs without decoration. If $\mathcal{G} <_g \mathcal{G}'$ then $(\mathcal{G}, \Lambda) <_g (\mathcal{G}', \Lambda')$.*

We fix notations for the concepts introduced above in the following definition.

DEFINITION 2.56. Let $(\mathcal{G}, \Lambda) \in \mathbf{G}$, we say that the order $\leq_c$ defined above is the *canonical order* of $(\mathcal{G}, \Lambda)$.

Let $\leq_a$ be an admissible ordering of $(\mathcal{G}, \Lambda)$. Then if $\leq_t$ is the total order defined in property 4 of the last proposition, we say that it is the *total ordering associated with* $\leq_a$.

An *ordered graph* is, by definition, a decorated graph with an admisible ordering.

CHAPTER 3

The geometry of the partition

In this chapter we will study the geometric properties of each of the subsets of the partition defined before. We prove that each of them has a natural structure of algebraic variety, as one can expect after the definition of the partition. We also study their geometric properties; here the results are quite surprising: the subsets of the partition are smooth and rational, and we can give a very detailed geometric description of each of the subsets of the partition in terms of the associated graph.

As a new ingredient, we will need a certain type of varieties that parametrize sets of infinitely near points with a given graph; they will be a fundamental technical tool during the rest of this memoir; it also will turn out that the subsets of the partition are fibered over them.

Our definition of behaviour graph plays a fundamental role in the way in which we obtain these results; if for the definition the partition we would have used embedded resolution graphs of the singularities at infinity instead of behaviour graphs, we only would be able to deduce the algebraicity of the subsets of the partition, but many of the results about the geometric structure could not be obtained.

3.1. Basic definitions

Convention: from now on when we say "graph" we will actually mean an $\mathcal{A}$-constructible graph.

NOTATION 3.1. The notation $(\mathcal{G}, \Lambda)$ for decorated graphs used before would yield quite heavy formulae in the rest of the paper, and hence we need to change it. We choose to denote a decorated graph by a single letter $\mathcal{G}$, and the graph obtained by deleting the decoration by $\mathcal{G}'$. If we need to mention the decoration of $\mathcal{G}$ we will denote it, as usually, by Λ.

DEFINITION 3.2. Let $(\mathcal{G}, \leq_a)$ be an ordered decorated graph. Define $V_{\mathcal{G},\leq_a}$ to be the set of pairs (I, α), where I is a s.s.i.n.p. to L and α an isomorphism from the graph of I to $\mathcal{G}$. The group of automorphisms $Aut(\mathcal{G})$ acts on $V_{\mathcal{G},\leq_a}$ by composition of the left. Define $V_{\mathcal{G}}$ to be quotient of $V_{\mathcal{G},\leq_a}$ by $Aut(\mathcal{G})$. We will denote the quotient mapping by

$$\kappa_{\mathcal{G},\leq_a} : V_{\mathcal{G},\leq_a} \to V_{\mathcal{G}}$$

As $\leq_a$ induces a total ordering on the graph $\mathcal{G}$ we can view any element (I, α) of $V_{\mathcal{G},\leq_a}$ as a model for the s.s.i.n.p. I. We have a canonical order in $\mathcal{G}$, that, in general, due to the decoration of $\mathcal{G}$, is stronger than the canonical order on $\mathcal{G}'$ (the graph of I). If $(I, [\alpha])$ is an element of $V_{\mathcal{G}}$ then the set I inherits, via any representative of $[\alpha]$, the canonical order of the set of vertices of $\mathcal{G}$. Hence we can view $(I, [\alpha])$ as a s.s.i.n.p. together with an extension of the parental order.

REMARK 3.3. If $\leq'_a$ is another admissible ordering in $\mathcal{G}$ then there exists a unique automorphism $\beta \in Aut(\mathcal{G})$ such that $\beta : (\mathcal{G}, \leq_a) \to (\mathcal{G}, \leq'_a)$ is an isomorphism of ordered graphs. Therefore given $\leq_a$ and $\leq'_a$, two admissible orderings for $\mathcal{G}$ there is a canonical bijection between the sets $V_{\mathcal{G},\leq_a}$ and $V_{\mathcal{G},\leq'_a}$.

DEFINITION 3.4. Given $(\mathcal{G}, \leq_a)$ an ordered decorated graph, a $(\mathcal{G}, \leq_a)$-*ordered polynomial* is a triple (f, I, α) where $f \in \mathbb{C}[x, y]$, the model (I, α) is a partial resolution of f, and α is an isomorphism from the decorated graph of the partial resolution to $(\mathcal{G}, \leq_a)$.

DEFINITION 3.5. Define $X_{\mathcal{G},\leq_a}$ to be the set of $(\mathcal{G}, \leq_a)$-ordered polynomials. Define $X_{\mathcal{G}}$ to be the set of pairs (f, I) where $I \subset I_f$ is a partial resolution set of f such that the graph of the partial resolution is isomorphic to $\mathcal{G}$. Define the "forgetting ordering" map

$$\rho_{\mathcal{G},\leq_a} : X_{\mathcal{G},\leq_a} \to X_{\mathcal{G}}$$

to be the map that takes (f, I, α) to (f, I).

REMARK 3.6. The group of automorphisms $Aut(\mathcal{G})$ acts on the left on $X_{\mathcal{G},\leq_a}$ and $X_{\mathcal{G}}$ might be viewed as the quotient by this action.

REMARK 3.7. Given $\leq_a$ and $\leq'_a$, two admissible orderings for $\mathcal{G}$ there is a canonical bijection between the sets $X_{\mathcal{G},\leq_a}$ and $X_{\mathcal{G},\leq'_a}$.

LEMMA 3.8. *Let $\mathcal{B}$ be a behaviour graph, then $X_{\mathcal{B}}$ can be identified with the set of polynomials such that $\mathcal{B}_f$ is isomorphic to $\mathcal{B}$, i.e. with the subset of the partition corresponding to $\mathcal{B}$.*

PROOF. Let $(f, I) \in X_{\mathcal{B}}$. The graph $\mathcal{G}_I$ of the partial resolution π is isomorphic to $\mathcal{B}$. As the decoration of $\mathcal{B}$ is such that Λ_2 and Λ_4 are empty, we conclude that I is the behaviour set of f. Hence I is determined by f. □

DEFINITION 3.9. We define the mappings

$$\tau_{\mathcal{G},\leq_a} : X_{\mathcal{G},\leq_a} \to V_{\mathcal{G},\leq_a} \quad \tau_{\mathcal{G}} : X_{\mathcal{G}} \to V_{\mathcal{G}}$$

by $\tau_{\mathcal{G},\leq_a}(f, I, \alpha) := (I, \alpha)$ and $\tau_{\mathcal{G}}(f, I) := (I, [\alpha])$ where α is any isomorphism from the graph of the partial resolution of f determined by I to $(\mathcal{G}, \leq_a)$.

We have to check that the definition does not depend on the chosen isomorphism α: if α' were a different choice then $\alpha' \circ \alpha^{-1}$ would be an automorphism of $\mathcal{G}$.

REMARK 3.10. The following commutativity relation holds

$$\kappa_{\mathcal{G},\leq_a} \circ \tau_{\mathcal{G},\leq_a} = \tau_{\mathcal{G}} \circ \rho_{\mathcal{G},\leq_a}$$

3.2. Varieties parametrizing infinitely near points

Given any graph $\mathcal{G}$, we will give natural structures of algebraic varieties to $V_{\mathcal{G},\leq_a}$ and $V_{\mathcal{G}}$, that make $\kappa_{\mathcal{G},\leq_a}$ become a morphism. We will also study the geometric properties of these varieties and of the morphism $\kappa_{\mathcal{G},\leq_a}$.

To fix notations we recall the following well known definition:

DEFINITION 3.11. Let S be an algebraic variety. A S-*variety* (X, p) consits of an algebraic variety X and a morphism $p : X \to S$, which will be called *structure morphism.* A S-variety (X, p) is said to be *smooth* if p is a smooth morphism. A S-*morphism* φ between two S-varieties (X_1, p_1), (X_2, p_2), is a morphism $\varphi : X_1 \to X_2$ that satisfies $p_2 \circ \varphi = p_1$.

NOTATION 3.12. When no confussion arises we will denote a S-variety (X, p) simply by X.

The following construction is due to S. Kleiman (see [**21**]):

CONSTRUCTION 3.13. Let X be any smooth surface, let $Bl_x X$ denote the variety obtained from the blow up of X at x. We are going to construct inductively a variety $X(n)$ parametrizing n-uples $(x_1, ..., x_n)$ with $x_i \in Bl_{x_{i-1}}...Bl_{x_1}X$, i.e. a variety parametrizing ordered sequences of n infinitely near points to X. We will also construct a universal family over it.

Step 1. Let $X(1) = X$; the universal family is constructed by considering $\pi_1(1)$, the blowing up at the diagonal of $X \times X$, as a X-morphism with via the second projection $p_2 : X \times X \to X$,

$$\begin{array}{ccc} Y & \stackrel{\pi_1(1)}{\longrightarrow} & X \times X \\ & \searrow \qquad \swarrow \; p_2 & \\ & X & \end{array}$$

The fibre of $\pi_1(1)_x$ over any $x \in X(1)$ is the blowing up of X at x.

Inductive step. Given k, a natural number, assume the existence of a smooth algebraic variety $X(k)$ together with:

- A bijection $\Psi(k)$ between the closed points of $X(k)$ and the set of k-uples $(x_1, ..., x_k)$ such that $x_i \in Bl_{x_{i-1}}...Bl_{x_1}X$
- A universal family, consisting of a sequence of $X(k)$-morphisms of algebraic smooth $X(k)$-varieties

$$Y_k(k) \stackrel{\pi_k(k)}{\longrightarrow} ... \stackrel{\pi_1(k)}{\longrightarrow} Y_0(k)$$

such that

(1) $Y_0(k) = (X \times X(k), p_2)$ where p_2 is the second projection.
(2) $\pi_i(k)$ is the blowing up at a smooth codimension 2 subvariety, and fibrewise over $x \in X(k)$, it is the blowing up $Bl_{x_i}...Bl_{x_1}X \to Bl_{x_{i-1}}...Bl_{x_1}X$ with center at x_i, where $(x_1, ..., x_k) = \Psi(k)(x)$.

Next we prove the existence of an algebraic variety $X(k+1)$, together with a bijection $\Psi(k+1)$ and a universal family with the properties stated above substituting k by $k+1$.

Define $X(k+1) := Y_k(k)$. Denote by $\beta_k : Y_k(k) \to X(k)$ the structure morphism of $Y_k(k)$. Notice that, if $\Psi(k)(x) = (x_1, ..., x_k)$, the fibre of β_k over $x \in X(k)$ is $Bl_{x_k}...Bl_{x_1}X$. If we define

$$\Psi(k+1)(x) := (x_1, ..., x_k, x) = (\Psi(k)(x), x)$$

we get a bijection between the closed points of $X(k+1)$ and the $k+1$-tuples $(x_1, ..., x_{k+1})$ such that $x_i \in Bl_{x_{i-1}}...Bl_{x_1}X$.

The sequence of morphisms of $X(k+1)$-varieties

$$Y_k(k+1) \stackrel{\pi_k(k+1)}{\longrightarrow} ... \stackrel{\pi_1(k+1)}{\longrightarrow} Y_0(k+1)$$

is defined to be the pullback of the sequence of morphisms of $X(k)$ varieties

$$Y_k(k) \stackrel{\pi_k(k)}{\longrightarrow} ... \stackrel{\pi_1(k)}{\longrightarrow} Y_0(k)$$

to $X(k+1)$ by the morphism $\beta_k : X(k+1) \to X(k)$.

Notice that then $Y_k(k+1) = Y_k(k) \times_{X(k)} Y_k(k)$, and that the diagonal Δ of this fibre product is a smooth codimension 2 subvariety: if we fibre the embedding $\Delta \subset Y_k(k+1) = Y_k(k) \times_{X(k)} Y_k(k)$ over the smooth variety $Y(k)$ via the second projection we find that each fibre is of the form $\{x\} \subset Bl_{x_k}...Bl_{x_1}X$ where $\{x\}$ is a point with reduced structure. Define

$$\pi_{k+1}(k+1) : Y_{k+1}(k+1) \to Y_k(k+1)$$

to be the blowing up of $Y_k(k+1) = Y_k(k) \times_{X_(k)} Y_k(k)$ at the diagonal. If $\Psi(k+1)(x) = (x_1, ..., x_{k+1})$, then $\pi_i(k+1)$, fibrewise over $x \in X(k+1)$, is the blowing up $Bl_{x_{k+1}}...Bl_{x_1}X \to Bl_{x_k}...Bl_{x_1}X$. This concludes the induction step and hence the construction.□

From now on we assume $X = \mathbb{P}^2$ and we let $L \subset \mathbb{P}^2$ be the line at infinity. Denote by $\mathbb{P}^2_L(n)$ the subset formed by the sequences of points infinitely near to L. Note that if $(\mathcal{G}, \leq_a)$ is a (possibly decorated) ordered graph with n black nodes then $V_{\mathcal{G},\leq_a} \subset \mathbb{P}^2(n)$. The following theorem have been proved independently by J. Roé in[**30**]. In the same paper, given any surface X, he considers the partition of $X(n)$ whose subsets consits of elements of $X(n)$ whose exceptional divisor has a given combinatorial structure; he shows that this partition is not neccesarily a stratification. When $X = \mathbb{P}^2$ the restriction of this partition to $\mathbb{P}^2_L(n)$ is the partition whose subsets are $V_{\mathcal{G},\leq_a}$ for all (possibly decorated) ordered graphs with n black nodes.

THEOREM 3.14. *Let $(\mathcal{G}, \leq_a)$ be an ordered graph. Let n be the number of vertices of $\mathcal{G}$ different from the root and d the number of free vertices. Then the set $V_{\mathcal{G},\leq_a}$ is a smooth locally closed subset of $\mathbb{P}^2_L(n)$ of dimension d. The pullback of the universal family*

$$Y_n(n) \stackrel{\pi_n(n)}{\longrightarrow} ... \stackrel{\pi_1(n)}{\longrightarrow} Y_0(n) = \mathbb{P}^2 \times \mathbb{P}^2(n)$$

to $V_{\mathcal{G},\leq_a}$ is a sequence of $V_{\mathcal{G},\leq_a}$-morphisms

$$Y_n(\mathcal{G}, \leq_a) \stackrel{\pi_n(\mathcal{G},\leq_a)}{\longrightarrow} ... \stackrel{\pi_1(\mathcal{G},\leq_a)}{\longrightarrow} Y_0(\mathcal{G}, \leq_a) = V_{\mathcal{G},\leq_a} \times \mathbb{P}^2$$

that forms a universal family over $V_{\mathcal{G},\leq_a}$. Each $\pi_i(\mathcal{G}, \leq_a)$ is a blowing up at a smooth center of codimension 2 in $Y_{i-1}(\mathcal{G}, \leq_a)$. Let $\pi(\mathcal{G}, \leq_a)$ be the composition $\pi_1(\mathcal{G}, \leq_a)\circ...\circ\pi_n(\mathcal{G}, \leq_a)$. We denote by $\mathcal{E}_i$ the exceptional divisor of $\pi_i(\mathcal{G}, \leq_a)$ and by $\mathcal{E}_0$ to the strict transform of the divisor $V_{\mathcal{G},\leq_a} \times L$. Then, the total transform of $V_{\mathcal{G},\leq_a} \times L$ by $\pi(\mathcal{G}, \leq_a)$ is a normal crossings divisor whose components are $\mathcal{E}_0$,...,$\mathcal{E}_n$. Moreover, if

$$\pi : X_n \stackrel{\pi_n}{\longrightarrow} ... \stackrel{\pi_1}{\longrightarrow} \mathbb{P}^2$$

is a blowing up process of $V_{\mathcal{G},\leq_a}$, then the fibre of $\mathcal{E}_i$ over π is exactly the exceptional divisor E_i for the blowing up process π.

PROOF. The proof is by induction over n. The step $n = 1$ is obvious and we leave it to the reader.

The canonical ordering in $\mathcal{G}$ together with the admissible ordering $\leq_a$ defines a total order $\leq_t$ in $\mathcal{G}$. Consider the last vertex x_n of $(\mathcal{G}, \leq_a)$ and let $\mathcal{G}_0$ be the graph obtained by the deletion of x_n. Denote by $\leq_{t,0}$ the total order that $\leq_t$ induces in $\mathcal{G}_0$, we have seen that there exists an admissible ordering $\leq_{a,0}$ in $\mathcal{G}_0$ such that $\leq_{t,0}$ is the total ordering obtained from the canonical ordering in $\mathcal{G}_0$ and the admissible ordering $\leq_{a,0}$. By induction hypothesis $V_{\mathcal{G}_0,\leq_{a,0}} \subset \mathbb{P}^2(n-1)$ is a smooth locally closed subset of $X(n-1)$ of dimension d_0, the number of *free* vertices of $\mathcal{G}_0$. The pullback of the universal family over $X(n-1)$ to $V_{\mathcal{G}_0,\leq_{a,0}}$ is a universal family

$$Y_{n-1}(\mathcal{G}_0,\leq_{a,0}) \overset{\pi_{n-1}(\mathcal{G}_0,\leq_{a,0})}{\longrightarrow} \ldots \overset{\pi_1(\mathcal{G}_0,\leq_{a,0})}{\longrightarrow} Y_0(\mathcal{G}_0,\leq_{a,0})$$

over $V_{\mathcal{G}_0,\leq_{a,0}}$. Moreover the total transform by $\pi(\mathcal{G}_0,\leq_{a,0})$ of $V_{\mathcal{G}_0,\leq_{a,0}} \times L$ is normal crossings with components $\mathcal{E}_{0,0}$,...,$\mathcal{E}_{n-1,0}$ that satisfy the statement that is made about them in the theorem. There are two different cases:

Case 1. The set $p(x_n)$ has only one element. Then $d_0 = d-1$. Let $p(x_n) = x_i$. Define

$$\dot{\mathcal{E}}_{i,0} := \mathcal{E}_{i,0} \setminus \bigcup_{j\neq i} \mathcal{E}_{i,0} \bigcap \mathcal{E}_{j,0}$$

From the properties of the divisor $\sum_{i=0}^{n-1} \mathcal{E}_{i,0}$ one deduces that the restriction to $\dot{\mathcal{E}}_{i,0}$ of the morphism $Y_{n-1}(\mathcal{G}_0,\leq'_a) \to V_{\mathcal{G}_0,\leq_{a,0}}$ is a smooth morphism of relative dimension 1. Moreover it is not hard to see that $V_{\mathcal{G},\leq_a} = \dot{\mathcal{E}}_{i,0}$. This shows that $V_{\mathcal{G},\leq_a}$ is a smooth locally closed subset of $Y_{n-1}(\mathcal{G}_0,\leq_{a,0})$ of dimension $d_0+1 = d$. The variety $Y_{n-1}(\mathcal{G}_0,\leq_{a,0})$ is the pullback of $Y_{n-1}(n-1) \to \mathbb{P}^2(n-1)$ to $V_{\mathcal{G}_0,\leq_{a,0}}$. As $V_{\mathcal{G}_0,\leq_{a,0}} \subset \mathbb{P}^2(n-1)$ is a locally closed inclusion we deduce that $Y_{n-1}(\mathcal{G}_0,\leq_{a,0})$ is locally closed in $Y_{n-1}(n-1) = \mathbb{P}^2(n)$. Hence $V_{\mathcal{G},\leq_a}$ is a smooth locally closed subset of $\mathbb{P}^2(n)$ of dimension d.

The pullback to $V_{\mathcal{G},\leq_a}$ of the universal family over $\mathbb{P}^2(n)$ gives a universal family over $V_{\mathcal{G},\leq_a}$. The first $n-1$ morphisms of the universal family are the pullback of the universal family over $V_{\mathcal{G}_0,\leq_{a,0}}$, this shows that all of them are blowing ups at smooth 2-codimensional centers and that the total transform of $V_{\mathcal{G},\leq_a} \times L$ by the composition of this $n-1$ blowing ups is the normal crossings divisor

$$\cup_{j=0}^{n-1} \mathcal{E}_{j,0} \times_{V_{\mathcal{G}_0,\leq_{a,0}}} V_{\mathcal{G},\leq_a}$$

Consider the last morphism of the universal family

$$Y_n(\mathcal{G},\leq_a) \overset{\pi_n(\mathcal{G},\leq_a)}{\longrightarrow} Y_{n-1}(\mathcal{G},\leq_a)$$

One verifies easily that the morphism $\pi_n(\mathcal{G},\leq_a)$ is the blowing up in at the diagonal of

$$\dot{\mathcal{E}}_{i,0} \times_{V_{\mathcal{G}_0,\leq_{a,0}}} \dot{\mathcal{E}}_{i,0} = \dot{\mathcal{E}}_{i,0} \times_{V_{\mathcal{G}_0,\leq_{a,0}}} V_{\mathcal{G},\leq_a}$$

As the diagonal is a 2-codimensional smooth center we deduce that $\pi_n(\mathcal{G},\leq_a)$ is a blowing up in a smooth codimension two center and that the total transform of $V_{\mathcal{G},\leq_a} \times L$ by $\pi(\mathcal{G},\leq_a)$ is a normal crossings divisor whose components are $\mathcal{E}_0$,...,$\mathcal{E}_n$. The rest of the statements about this divisors can be checked easily at the level of points.

Case 2. The set $p(x_n)$ has two elements. Then $d_0 = d$. Let $p(x_n) = \{x_i, x_j\}$. In this case $V_{\mathcal{G},\leq_a} = \mathcal{E}_i \bigcap \mathcal{E}_j \cong V_{\mathcal{G}_0,\leq_{a,0}}$. We leave the rest to the reader. □

The following is a corollary of the proof of the last theorem:

COROLLARY 3.15. *Let $(\mathcal{G},\leq_a)$ be an ordered graph with n vertices without counting the root and d free vertices. For any $i \geq 1$. Let $(\mathcal{G}_i,\leq_{a,i})$ denote the ordered graph obtained from $(\mathcal{G},\leq_a)$ by deletion of the last $n-i$ vertices. The sequence of morphisms*

$$V_{\mathcal{G}_n,\leq_{a,n}} \overset{\phi_n}{\longrightarrow} \ldots \overset{\phi_1}{\longrightarrow} V_{\mathcal{G}_0,\leq_{a,0}} = Spec(\mathbb{C})$$

defined by the property that each ϕ_i "forgets the i-th infinitely near point" satisfies that

- *If the i-th vertex is free then each fibre of ϕ_i is $\mathbb{P}^1$ minus a finite number of points that depends only on the graph $\mathcal{G}$.*
- *If the i-th vertex is fixed then ϕ_i is an isomorphism.*

Moreover $V_{\mathcal{G}_1,\leq_{a,1}}$ can be identified with L and hence it is isomorphic to $\mathbb{P}^1$.

REMARK 3.16. Consider an ordered graph $(\mathcal{G}, \leq_a)$ and let $v_0, ..., v_n$ be the set of vertices with the order that $\leq_a$ induces. For each $i > 0$, let $v_{m(i)}$ be the smallest vertex of $p(v_i)$. Consider any ordered set of infinitely near points $(x_1, ..., x_n) \in V_{\mathcal{G},\leq_a}$, and let $X_n \to ... \to \mathbb{P}^2$ be the associated blowing up process. For any h such that $m(i) \leq h < i$ the infinitely near point x_i has a representative in X_h, which we simply denote by x_i. We define for each h a section of $\pi_1(\mathcal{G}, \leq_a) \circ ... \circ \pi_h(\mathcal{G}, \leq_a)$

$$s_{i,h} : V_{\mathcal{G},\leq_a} \to Y_h(\mathcal{G}, \leq_a)$$

that takes each $(x_1, ..., x_n) \in V_{\mathcal{G},\leq_a}$ to $x_i \in Y_h(\mathcal{G}, \leq_a)$. These sections are algebraic.

Next we study the mapping $\kappa_{\mathcal{G},\leq_a} : V_{\mathcal{G},\leq_a} \to V_{\mathcal{G}}$.

NOTATION 3.17. Given a graph $\mathcal{G}$, the set of $\sim_c$-equivalence classes is totally ordered by the canonical order. Let r be the number of equivalence classes, n_i the number of vertices of the i-th equivalence class and $m_i := \sum_{j=1}^{i} n_j$. Fix an admissible ordering $\leq_a$ on $\mathcal{G}$, and let $\leq_t$ be the induced total ordering. Consider $(I, [\alpha]) \in V_{\mathcal{G}}$. Fixing a representative α of $[\alpha]$ we fix a model for I and hence an ordered sequence $(\pi_1, ..., \pi_{m_r})$ of blowing ups of points. Define

$$\pi_i'(\mathcal{G}) := \pi_{m_{i-1}+1}(\mathcal{G}, \leq_a) \circ ... \circ \pi_{m_i}(\mathcal{G}, \leq_a) : Y_{m_i}(\mathcal{G}, \leq_a) \to Y_{m_i-1}(\mathcal{G}, \leq_a) \tag{3.1}$$

The next remark implies that if we change α by a different isomorphism within its class then the sequence of morphisms defined above will differ by a canonical isomorphism. Hence we do not include $\leq_a$ in the notation $\pi_i'(\mathcal{G})$.

REMARK 3.18. Let X be a smooth surface, and p, q two different points on it. Then $Bl_q Bl_p X$ is canonically isomorphic to $Bl_p Bl_q X$.

NOTATION 3.19. Let $f : X \to Y$ be a morphism, β (resp. β') a bijection from X to itself (resp. from Y to itself) such that $\beta' \circ f = f \circ \beta$. Then given $y \in Y$ the automorphism β defines an bijection between X_y and $X_{\beta'(y)}$ (where X_y denotes the fibre of f over y). We denote this bijection by β_y.

The group $Aut(\mathcal{G})$ acts on $V_{\mathcal{G},\leq_a}$, this action can be lifted at the level of points to each $Y_{m_i}(\mathcal{G}, \leq_a)$ as follows: consider $\beta \in Aut(\mathcal{G})$ and $(I, \alpha) \in V_{\mathcal{G},\leq_a}$. Denote by h the structure morphism $h : Y_{m_i}(\mathcal{G}, \leq_a) \to V_{\mathcal{G},\leq_a}$. The automorphism β defines bijection from $V_{\mathcal{G},\leq_a}$ to itself. We define a bijection, that we also call β, from $Y_{m_i}(\mathcal{G}, \leq_a)$ to itself, by imposing that $\beta \circ h = h \circ \beta$ and letting

$$\beta_{(I,\alpha)} : Y_{m_i}(\mathcal{G}, \leq_a)_{(I,\alpha)} \to Y_{m_i}(\mathcal{G}, \leq_a)_{(I,\beta\circ\alpha)}$$

be the canonical isomorphism that exists due to the fact that both $Y_{m_i}(\mathcal{G}, \leq_a)_{(I,\alpha)}$ and $Y_{m_i}(\mathcal{G}, \leq_a)_{(I,\beta\circ\alpha)}$ are obtained by blowing up the same set of infinitely near points in a different order.

Notice that the lifting defined above is such that the sequence

$$Y_{m_r}(\mathcal{G}, \leq_a) \overset{\pi_r'(\mathcal{G})}{\longrightarrow} ... \overset{\pi_1'(\mathcal{G})}{\longrightarrow} Y_0(\mathcal{G}, \leq_a)$$

has the property that the morphisms are compatible the action by $Aut(\mathcal{G})$.

Define $\mathcal{G}_i$ to be the graph obtained by deleting all the vertices whose $\sim_c$-class is not among the first i $\sim_c$-classes. By abuse of language denote by $\leq_a$ the restriction of the admissible ordering $\leq_a$ to $\mathcal{G}_i$. Notice that as $\mathcal{G}_i$ is left invariant by the action of any element of $Aut(\mathcal{G}_{i+1})$. Restriction to $\mathcal{G}_i$ defines an homomorphism

$$\rho_{i+1} : Aut(\mathcal{G}_{i+1}) \rightarrow Aut(\mathcal{G}_i)$$

We denote by K_{i+1} the kernel of ρ_{i+1}. The admissible ordering $\leq_a$ determines a section

$$\sigma_{i+1} : im(\rho_{i+1}) :\longrightarrow Aut(\mathcal{G}_{i+1})$$

of ρ_{i+1} as follows:

Let $\beta : \mathcal{G}_i \rightarrow \mathcal{G}_i$ be an automorphism in $im(\rho_{i+1})$. If b is a vertex of $\mathcal{G}_{i+1}$ that does not belong to the $i+1$-th $\sim_c$-class then define $\sigma_{i+1}(\beta)(b) := \beta(b)$. Assume that b is in the $i+1$-th $\sim_c$-class. The set $[b] \bigcap p^{-1}(p(b))$ (where p is the parental function) is a set that is totally ordered by $\leq_a$ and contains b; let b the j-th element of $p^{-1}(p(b))$. Assume that $p(b)$ is a set of one element and denote this element by b'; as b' belongs to $\mathcal{G}_i$ then $b'' := \beta(b')$ is well defined. The facts that $b' \sim_c b''$ and that $\beta \in im(\rho_{i+1})$ implies that $[b] \bigcap p^{-1}(b'')$ is a set totally ordered by $\leq_a$ with the same number of elements than $[b] \bigcap p^{-1}(b')$. Define $\sigma_{i+1}(\beta)(b)$ to be the j-th element of $[b] \bigcap p^{-1}(b'')$. The definition of $\sigma_{i+1}(\beta)(b)$ in the case in which $p(b)$ has two elements follows the same pattern. It is easy to check that σ_{i+1} is a homomorphism and a section of ρ_{i+1}.

The existence of the section ρ_{i+1} imply that $Aut(\mathcal{G}_{i+1})$ *is the semidirect product of* $im(\rho_{i+1})$ *and* K_{i+1}.

Let $[b]$, the $i+1$-th $\sim_c$-class of $\mathcal{G}_{i+1}$ and consider the set $S := \{p(b) : b \in [b]\}$. All the fibres of the mapping $p_{|[b]} : [b] \rightarrow S$ are finite sets with the same number of elements. It is easy to see that K_{i+1} is the direct product of the permutation groups of the fibres of p:

$$K_{i+1} = \prod_{s \in S} \mathcal{S}_{p_{|[b]}^{-1}(s)}.$$

PROPOSITION 3.20. *The group $Aut(\mathcal{G})$ acts algebraically on $V_{\mathcal{G},\leq_a}$, moreover the lifting of the action to the sequence of morphisms*

$$Y_{m_r}(\mathcal{G}, \leq_a) \stackrel{\pi'_r(\mathcal{G})}{\longrightarrow} \ldots \stackrel{\pi'_1(\mathcal{G})}{\longrightarrow} Y_0(\mathcal{G}, \leq_a)$$

is also algebraic.

PROOF. We show the proposition by induction on r the number of $\sim_c$-classes of $\mathcal{G}$. We can start the induction by $i = 0$ for which there is nothing to prove. For the inductive step let $(\mathcal{G}, \leq_a)$ be a graph with r $\sim_c$-classes. Let $[b_1]$ be the $r-th$ $\sim_c$-class. We are going to consider two cases

Case 1. The set $p(b_1)$ is a single vertex. Consider the graph $(\mathcal{G}_{r-1}, \leq_a)$. As it has $r-1$ $\sim_c$-classes we can assume by induction that $Aut(\mathcal{G}_{r-1})$ acts algebraically on $V_{\mathcal{G}_{r-1},\leq_a}$ and that there is an algebraic lifting of the action to

$$Y_{m_{r-1}}(\mathcal{G}_{r-1}, \leq_a) \stackrel{\pi'_{r-1}(\mathcal{G}_{r-1})}{\longrightarrow} \ldots \stackrel{\pi'_1(\mathcal{G}_{r-1})}{\longrightarrow} Y_0(\mathcal{G}_{r-1}, \leq_a).$$

Let $p(b_1) = b'_1$. The mapping $p : [b_1] \rightarrow [b'_1]$ is a surjection such that all the fibres have the same number of elements. Let this number be t. Let $[b'_1] = \{b'_1, ..., b'_s\}$. For each $i \in \{1, ..., s\}$ there exists a divisor $\mathcal{E}_i$ such that, its fibre over each $(I,\alpha) \in V_{\mathcal{G}_{r-1},\leq_a}$ is the exceptional divisor of the blowing up of the infinitely near point of I represented by b_i'. One checks at the level of points that the set

$$\bigcup_{i=1}^{s} \dot{\mathcal{E}}_i$$

is a $im(\rho_r)$-invariant locally closed subset. Hence $im(\rho_r)$ acts algebraically on it. This implies that $im(\rho_r)$ acts algebraically on the variety

$$V := \prod_{i=1}^{s} (\underbrace{\dot{\mathcal{E}}_i \times_{V_{\mathcal{G}_{r-1},\leq_a}} \dots \times_{V_{\mathcal{G}_{r-1},\leq_a}} \dot{\mathcal{E}}_i}_{t\ times}) \setminus \Delta_i)$$

where Δ_i consists of the set of t-uples with at least 2 coordinates repeated. Examining the proof of Theorem 3.14 it is not difficult to conclude that $V = V_{\mathcal{G},\leq_a}$. Hence $im(\rho_r)$ acts algebraically on $V_{\mathcal{G},\leq_a}$. Notice that an element of V is an s-uple $(y_1, \dots, y_s)$ where each y_i is a t-uple

$$(z_1, \dots, z_t) \in (\underbrace{\dot{\mathcal{E}}_i \times_{V_{\mathcal{G}_{r-1},\leq_a}} \dots \times_{V_{\mathcal{G}_{r-1},\leq_a}} \dot{\mathcal{E}}_i}_{t\ times}) \setminus \Delta_i$$

As

$$K_r = \prod_{i=1}^{s} \mathcal{S}_t$$

and an element of K_r acts on any $(y_1, \dots, y_s)$ by permutation of the entries of each t-uple y_i we can conclude that K_r acts algebraically on $V_{\mathcal{G},\leq_a}$. As $Aut(\mathcal{G})$ is the semidirect product of $im(\rho_r)$ and K_r, and both of them act algebraically on $V_{\mathcal{G},\leq_a}$, we can conclude that the action of $Aut(\mathcal{G})$ on $V_{\mathcal{G},\leq_a}$ is algebraic.

It remains to show that the lifting of the action to $Y_i(\mathcal{G}, \leq_a)$ is algebraic for any $i \in \{1, \dots, r\}$. Arguments similar to the proof of Theorem 3.14 show that for any $i \leq r-1$ we have an equality

$$Y_{m_i}(\mathcal{G}, \leq_a) = Y_{m_i}(\mathcal{G}_{r-1} \leq_a) \times_{V_{\mathcal{G}_{r-1},\leq_a}} V_{\mathcal{G},\leq_a}.$$

The action of $Aut(\mathcal{G})$ on $Y_{m_i}(\mathcal{G}, \leq_a)$ arises from the action of $Aut(\mathcal{G})$ on each of the three varieties involved in the fibre product. (This can be checked at the level of points). We have seen that the action of $Aut(\mathcal{G})$ on $V_{\mathcal{G},\leq_a}$ is algebraic. Moreover it is easy to check that the action of $Aut(\mathcal{G})$ on the varieties $Y_{m_i}(\mathcal{G}_{r-1}, \leq_a)$ and $V_{\mathcal{G}_{r-1},\leq_a}$ can be factorized via the homomorphism $\rho_r : Aut(\mathcal{G}) \to Aut(\mathcal{G}_{r-1})$. As the action of the last group on both varieties is algebraic we conclude that the action of $Aut(\mathcal{G})$ also. It follows that the action of $Aut(\mathcal{G})$ on $Y_{m_i}(\mathcal{G}, \leq_a)$ is algebraic for $i \leq r-1$.

We study the action of $Aut(\mathcal{G})$ on $Y_{m_r}(\mathcal{G}, \leq_a)$. The last variety arises from the blowing up of $Y_{m_{r-1}}(\mathcal{G}, \leq_a)$ at the smooth center Z defined as follows: make the identification

$$Y_{m_{r-1}}(\mathcal{G}, \leq_a) = Y_{m_{r-1}}(\mathcal{G}_{r-1}, \leq_a) \times_{V_{\mathcal{G}_{r-1},\leq_a}} V_{\mathcal{G},\leq_a}$$

and consider the second projection

$$p_2 : Y_{m_{r-1}}(\mathcal{G}_{r-1}, \leq_a) \times_{V_{\mathcal{G}_{r-1},\leq_a}} V_{\mathcal{G},\leq_a} \to V_{\mathcal{G},\leq_a}.$$

Z is defined to be the subscheme of $Y_{m_{r-1}}(\mathcal{G}, \leq_a)$ characterized by the property that the fibre of the restriction $p_2 : Z \to V_{\mathcal{G},\leq_a}$ over the point $(y_1, ..., y_s)$ is equal to the reduced 0-dimensional subscheme

$$\{z_{i,j} \ : \ 1 \leq i \leq s \ , \ 1 \leq j \leq t\}$$

where $y_i = (z_{i,1}, ..., z_{i,t})$. The scheme Z is a smooth subscheme of codimension 2. One can check at the level of points that $Aut(\mathcal{G})$ leaves it invariant. Hence the action of $Aut(\mathcal{G})$ on $Y_{m_{r-1}}(\mathcal{G}, \leq_a)$ lifts to an algebraic action on its blow up at Z, i.e. an algebraic action on $Y_{m_r}(\mathcal{G}, \leq_a)$. It is not difficult to check that this last action coincides at the level of points with our original action of $\mathcal{G}$ on $Y_{m_r}(\mathcal{G}, \leq_a)$. This concludes the proof of case 1.

Case 2. The set $p(b_1)$ has two vertices. The proof of this case is simpler than the proof of case 1 and does not involve new ideas. We leave it to the reader. □

Quotienting both $V_{\mathcal{G},\leq_a}$ and

$$Y_{m_r}(\mathcal{G}, \leq_a) \overset{\pi'_r(\mathcal{G})}{\longrightarrow} ... \overset{\pi'_1(\mathcal{G})}{\longrightarrow} Y_0(\mathcal{G}, \leq_a)$$

by $Aut(\mathcal{G})$, and noticing that the action of $Aut(\mathcal{G})$ on any of the varieties above is free we obtain:

THEOREM 3.21. *Let $\mathcal{G}$ be a (possibly decorated) graph. Assume that $\mathcal{G}$ has d free vertices. Then the set $V_{\mathcal{G}}$ has a natural structure of algebraic variety of dimension d. Moreover, if r is the number of $\sim_c$-equivalence classes of vertices there is a universal family of morphisms over $V_{\mathcal{G}}$*

$$\mathcal{Y}_r(\mathcal{G}) \overset{\pi''_r(\mathcal{G})}{\longrightarrow} ... \overset{\pi''_1(\mathcal{G})}{\longrightarrow} \mathcal{Y}_0(\mathcal{G})$$

such that, fibrewise over $x \in V_{\mathcal{G}}$, the morphism $\pi''_i(\mathcal{G})$ is the blowing up

$$Bl_{x_{m_i}}...Bl_{x_1}P^2 \to Bl_{x_{m_{i-1}}}...Bl_{x_1}\mathbb{P}^2$$

where $\{x_{m_{i-1}+1}, ..., x_{m_i}\}$ are the infinitely near points corresponding to the vertices of the i-th $\sim_c$-class of $\mathcal{G}$.

Moreover, if $\leq_a$ is an admissible ordering for $\mathcal{G}$ then $V_{\mathcal{G}}$, together with its universal family are the result of quotienting $V_{\mathcal{G},\leq_a}$ and the sequence of morphisms

$$Y_{m_r}(\mathcal{G}, \leq_a) \overset{\pi'_r(\mathcal{G})}{\longrightarrow} ... \overset{\pi'_1(\mathcal{G})}{\longrightarrow} Y_0(\mathcal{G}, \leq_a)$$

by $Aut(\mathcal{G})$. In particular the "forgetting order" map $\kappa_{\mathcal{G},\leq_a} : V_{\mathcal{G},\leq_a} \to V_{\mathcal{G}}$ is an algebraic covering morphism.

3.3. The algebraicity of the subsets of the partition

Given any behaviour graph $\mathcal{B}$ we prove that $X_{\mathcal{B}}$ is algebraic, we will also give an algebraic structure to $X_{\mathcal{B},\leq_a}$ in such a way that the mapping $\tau_{\mathcal{B},\leq_a}$ is a morphism.

LEMMA 3.22. *Let $\mathcal{G}, \leq_a$ be an ordered graph with n vertices. For $i > 1$ consider the i-th vertex v_i. Let the number $m(i)$ be defined as in remark 3.16. Consider the morphism*

$$\phi : V_{\mathcal{B},\leq_a} \to L$$

that takes each ordered set of infinitely near points $(x_1, ..., x_n)$ to its first point x_1. Then for any point p at infinity of $\mathbb{P}^2$ and any h such that $m(i) \leq h \leq i-1$ there exist $(\mathcal{U}, (u, v))$, where $\mathcal{U}$ is an open subset of $Y_h(\mathcal{G}, \leq_a)$ and u, v are two regular functions on it with the following property:

P(i,h): *Let (I,α) be any element of $\phi^{-1}(L\setminus\{p\})$. Consider the blowing up process*

$$\sigma_h : X_h \xrightarrow{\pi_h} \ldots \xrightarrow{\pi_1} \mathbb{P}^2$$

that consists in the first h blowing ups of (I,α); let $\mathcal{U}_{(I,\alpha)}$ be the fibre over (I,α) of the restriction to $\mathcal{U}$ of the composition $pr_1 \circ \pi_1(\mathcal{G},\leq_a)\circ\ldots\circ\pi_h(\mathcal{G},\leq_a)$, where $pr_1 : Y_0(\mathcal{G},\leq_a) = V_{\mathcal{G},\leq_a}\times\mathbb{P}^2$ is the first projection; let $u_{(I,\alpha)}, v_{(I,\alpha)}$ be the restrictions of (u,v) to the fibre. Then

$$(\mathcal{U}_{(I,\alpha)},(u_{(I,\alpha)},v_{(I,\alpha)}))$$

is an chart adapted to σ_h centered in x_i, the i-th point of (I,α).

PROOF. We will work with $p = (1:0:0)$, the general case is not more difficult. We will work by induction on i. If $i=1$ then $p(v_i)=w$. Let $(X:Y:Z)$ be the projective reference of $\mathbb{P}^2$, consider the affine chart $(U_X,(y,z))$ where $U_X = \mathbb{P}^2\setminus\{X=0\}$ and $(y,z)=(Y/X,Z/X)$. Given any $(I,\alpha)\in\phi^{-1}(L\setminus\{(1:0:0)\}$, the point x_1 (the first point of (I,α)) belongs to $L\setminus\{(1:0:0)\}$; we denote by $y(x_1)$ the first coordiante of x_1 respect to the chart $(U_X,(y,z))$. We can view $y(x_1)$ as a regular function on $\phi^{-1}(L\setminus\{(1:0:0)\}$. We define $(\mathcal{U},(u,v))$ as follows:

$$\mathcal{U} := \phi^{-1}(L\setminus\{(1:0:0)\})\times U_X$$

and, if pr_2 is the second projection of the last product, $u := pr_2^* y - y(x_1)$ and $v := pr_2^* z$.

Lets assume that the proposition is true for values smaller than $i-1$. Lets prove it for i. For simplicity we will work only in the case in which $h=m(i)$. With this assumption v_h is the last vertex of $p(v_i)$. By induction there exists $(\mathcal{U}',(u',v'))$ that satisfy the property **P**(h,h-1). For any i denote by $\sigma_i(\mathcal{G},\leq_a)$ the composition $\pi_1(\mathcal{G},\leq_a)\circ\ldots\circ\pi_i(\mathcal{G},\leq_a)$. Let φ be the composition $pr_1\circ\sigma_{h-1}(\mathcal{G},\leq_a)$, where pr_1 is the projection to the first factor of the product $Y_0(\mathcal{G},\leq_a)=V_{\mathcal{G},\leq_a}\times\mathbb{P}^2$. The locus of $\mathcal{U}'$ where both u' and v' vanish is the intersection of $(\phi\circ\varphi)^{-1}(L\setminus\{(1:0:0)\}$ with the center of the blowing up

$$\pi_h(\mathcal{G},\leq_a) : Y_h(\mathcal{G},\leq_a)\to Y_{h-1}(\mathcal{G},\leq_a).$$

It is easy to show that

$$((u',v'),\varphi) : \mathcal{U}'\to\mathbb{C}^2\times\phi^{-1}(L\setminus\{(1:0:0)\})$$

is an isomorphism. Using this we see that we can cover $\pi_h(\mathcal{G},\leq_a)^{-1}(\mathcal{U}')$ by two open subsets $\mathcal{U}_1$ and $\mathcal{U}_2$ such that there exists functions u_1, v_1 (resp. u_2, v_2) such that $(\mathcal{U}_1,(u_1,v_1))$ (resp. $(\mathcal{U}_2,(u_2,v_2))$ is, fibrewise, the first (resp. second) blowing up chart of the blowing up of $\mathcal{U}'$ at the locus where $\{u=v=0\}$. There are two cases:

Case 1 The set $p(v_i)$ has two vertices. In this case $p(v_i)=\{v_h,v_k\}$ with $v_k\geq_p v_h$. We choose $(\mathcal{U},(u,v))$ as follows:

- If v_k is the biggest vertex of $p(v_h)$ in the parental order (this includes the case in which $p(v_h)$ has only one vertex) then take $(\mathcal{U},(u,v))$ to be equal to $(\mathcal{U}_1,(u_1,v_1))$.
- If v_k is not the biggest vertex of $p(v_h)$ then take $(\mathcal{U},(u,v))$ to be equal to $(\mathcal{U}_2,(v_2,u_2))$ (notice the permutation in the order of the functions).

The reason of the dichotomy and of the permutation of the order of the functions in the second item is that, by definition 1.4, in a chart adapted to a blowing up process the oldest exceptional divisor meeting the center of the chart should be defined by the vanishing of the second coordinate function.

Case 2 The set $p(v_i)$ consists only on v_h. Then the image of the section $s_{i,h}$ (see Remark 3.16 for its definition) should lie in $\dot{\mathcal{E}}_h$, and hence the image of the restriction of $s_{i,h}$ to $L \setminus \{(1:0:0)\}$ lies in the locus of $\mathcal{U}_2$ where v_2 vanishes. We can choose $\mathcal{U} = \mathcal{U}_2$, $v = v_2$ and $u = u_2 - u_2 \circ s_{i,h} \circ pr_1 \circ \sigma_h(\mathcal{G}, \leq_a)$ This finishes the proof. □

In the following construction we introduce some ideal sheaves that will be quite useful to the study of $X_{\mathcal{B},\leq_a}$ for a behaviour graph $\mathcal{B}$.

CONSTRUCTION 3.23. For this construction we will use the following notations: consider $\mathcal{B}$ a behaviour graph with $n+1$ vertices, $\leq_a$ an admissible ordering on it and (I,α) an element of $V_{\mathcal{B},\leq_a}$. Let

$$\pi : X_n \xrightarrow{\pi_n} \dots \xrightarrow{\pi_1} \mathbb{P}^2$$

be the blowing up process that (I,α) defines and denote by x_i the center of π_i. Let σ_i be the composition $\pi_1 \circ \dots \pi_i$. We define sheaves of ideals of $\mathcal{O}_{\mathbb{P}^2}$ for each vertex v_i different from the root in the following way:

Case 1 The set $p(v_i)$ has only one element. In this case the arrow that we obtain when we delete v_i belongs to Λ_2; let (t,t',t'') be its weight. We let $p(v_i) = v_j$, then $x_i \in \dot{E}_j$. Consider (x,y) the coordinates of a chart of X_{i-1} adapted to σ_{i-1} and centered in x_i (this chart is such that E_j is defined by $y=0$).

- We define $\mathcal{I}_{I,i}$ to be the sheaf of functions f regular at $\sigma_{i-1}(x_i)$ such that if $\sigma_{i-1}^* f = \sum a_{i,j} x^i y^j$ respect to the coordinates (x,y) then

$$\sigma_{i-1}^* f \in y^{\mathrm{mtot}(v_j)} (x,y)^{t'}$$

$$a_{i,\mathrm{mtot}(v_j)} = 0 \quad \text{if} \quad i \leq t-1$$

$$a_{i,j+\mathrm{mtot}(v_j)} = 0 \quad \text{if} \quad i+j = t' \quad \text{and} \quad j \leq t''-1$$

- Define $J_{I,i,1}$ substituting the second condition imposed for defining $\mathcal{I}_i$ by

$$a_{i,\mathrm{mtot}(v_j)} = 0 \quad \text{if} \quad i \leq t$$

- Define $J_{I,i,2}$ substituting the third condition imposed for defining $\mathcal{I}_i$ by

$$a_{i,j+\mathrm{mtot}(v_j)} = 0 \quad \text{if} \quad i+j = t' \quad \text{and} \quad j \leq t''$$

Case 2 The set $p(v_i)$ has two elements. In this case the arrow that we obtain when we delete v_i belongs to Λ_4, let (t_1,t_2,t_3,t_4,t_5) be its weight. We let $p(v_i) = \{v_j, v_k\}$ with $v_j \geq_p v_k$, then $x_i = E_j \cap E_k$. Consider (x,y) the coordinates of a chart of X_{i-1} adapted to σ_{i-1} and centered in x_i (this chart is such that E_j is defined by $y=0$ and E_k by $x=0$).

- We define $\mathcal{I}_{I,i}$ to be the sheaf of functions f regular at $\sigma_{i-1}(x_i)$ such that if $\sigma_{i-1}^* f = \sum a_{i,j} x^i y^j$ with respect to the coordinates (x,y) then

$$\sigma_{i-1}^* f \in x^{\mathrm{mtot}(v_k)} y^{\mathrm{mtot}(v_j)} (x,y)^{t_3}$$

$$a_{\mathrm{mtot}(v_k)+i,\mathrm{mtot}(v_j)} = 0 \quad \text{if} \quad i \leq t_1 - 1$$

$$a_{\mathrm{mtot}(v_k),\mathrm{mtot}(v_j)+j} = 0 \quad \text{if} \quad j \leq t_2 - 1$$

$$a_{\mathrm{mtot}(v_k)+i,\mathrm{mtot}(v_j)+j} = 0 \quad \text{if} \quad i+j = t_3 \quad \text{and} \quad j \leq t_4 - 1$$

$$a_{\mathrm{mtot}(v_k)+i,\mathrm{mtot}(v_j)+j} = 0 \quad \text{if} \quad i+j = t_3 \quad \text{and} \quad i \leq t_5 - 1$$

- Define $J_{I,i,1}$ substituting the second condition imposed for defining $\mathcal{I}_i$ by
$$a_{\mathrm{mtot}(v_k)+i,\mathrm{mtot}(v_j)} = 0 \quad \text{if} \quad i \leq t_1$$
- Define $J_{I,i,2}$ substituting the third condition imposed for defining $\mathcal{I}_i$ by
$$a_{\mathrm{mtot}(v_k),\mathrm{mtot}(v_j)+j} = 0 \quad \text{if} \quad j \leq t_2$$
- Define $J_{I,i,3}$ substituting the fourth condition imposed for defining $\mathcal{I}_i$ by
$$a_{\mathrm{mtot}(v_k)+i,\mathrm{mtot}(v_j)+j} = 0 \quad \text{if} \quad i+j = t_3 \quad \text{and} \quad j \leq t_4$$
- Define $J_{I,i,4}$ substituting the fifth condition imposed for defining $\mathcal{I}_i$ by
$$a_{\mathrm{mtot}(v_k)+i,\mathrm{mtot}(v_j)+j} = 0 \quad \text{if} \quad i+j = t_3 \quad \text{and} \quad i \leq t_5$$

DEFINITION 3.24. Let $\mathcal{B}$ be a behaviour as considered above, and $(I, \alpha) \in V_{\mathcal{B},\leq_a}$. We define
$$\mathcal{I}_I := \cap_{i=1}^n \mathcal{I}_{I,i}.$$
For each $J_{I,i,j}$ (see the construction above for its definition) we define the ideal
$$\mathcal{J}_{I,i,j} := \mathcal{I}_I \cap J_{I,i,j}.$$

Since the ideal $\mathcal{I}_I$ above does not depend on α it is reasonable not to include it in the notation. However, the ideals $J_{I,i,j}$ depend on α, but what happens is that, changing α produces permutation of the different $\mathcal{J}_{I,i,j}$'s. As in the applications in which they appear we will use all of them together, the ambiguity introduced by not writing α in the notation is not so important, and on the other hand we will get lighter formulae.

LEMMA 3.25. *Let $\mathcal{B}$ be a behaviour graph and $\leq_a$ be an admissible ordering on it. Let $d = \deg(\mathcal{B})$. Consider $(I, \alpha) \in V_{\mathcal{B},\leq_a}$. Then the set of polynomials f such that (f, I, α) belongs to $X_{\mathcal{B},\leq_a}$ is equal to*

$$H^0(\mathbb{P}^2, \mathcal{I}_I(d)) \setminus \cup_{i,j} H^0(\mathbb{P}^2, \mathcal{J}_{I,i,j}(d)) \tag{3.2}$$

where i, j ranges over all the values for which $\mathcal{J}_{I,i,j}$ has been defined. Each of the $H^0(\mathbb{P}^2, \mathcal{J}_{I,i,j}(d))$ is of codimension at most 1 in $H^0(\mathbb{P}^2, \mathcal{I}_I(d))$.

PROOF. The proof of the first assertion is straight forward because the ideals have been constructed so that it works (see Proposition 2.49). For the second assertion notice that for any i, j

$$\mathrm{length}(\mathcal{I}_{I,i}/J_{I,i,j}) = 1 \tag{3.3}$$

because we define $J_{I,i,j}$ exactly by adding one more linear condition. This implies that

$$\mathrm{length}(\mathcal{I}_I/\mathcal{J}_{I,i,j}) \leq 1 \tag{3.4}$$

and this implies the assertion. □

PROPOSITION 3.26. *Let $d = \deg(\mathcal{B})$. Let $\mathcal{P}(d)$ be the set of polynomials in three variables that are homogeneous of degree d. The inclusion*

$$X_{\mathcal{B},\leq_a} \subset \mathcal{P}(d) \times V_{\mathcal{B},\leq_a} \tag{3.5}$$

is locally closed. Therefore $X_{\mathcal{B},\leq_a}$ inherits a structure of algebraic variety such that $\tau_{\mathcal{B},\leq_a}$ is a morphism whose fibres are either empty or complements of hyperplane arrangements in a vector space.

PROOF. If we show locally closedness, the others statements follow easily. For any i, j define the sets

$$\mathbf{H}(\mathcal{I}_i) := \cup_{(I,\alpha)\in V_{\mathcal{B},\leq_a}} H^0(\mathbb{P}^2, \mathcal{I}_{I,i}(d)) \times \{(I,\alpha)\}$$

$$\mathbf{H}(J_{i,j}) := \cup_{(I,\alpha)\in V_{\mathcal{B},\leq_a}} H^0(\mathbb{P}^2, J_{I,i,j}(d)) \times \{(I,\alpha)\}$$

and view them as sets fibered over $V_{\mathcal{B},\leq_a}$ by projection to the second component. We claim that they are all closed in $\mathcal{P}(d) \times V_{\mathcal{B},\leq_a}$. If the claim were true then the inclusion 3.5 would be locally closed as

$$X_{\mathcal{B},\leq_a} = (\cap_i \mathbf{H}(\mathcal{I}_i)) \setminus (\cup_{i,j} \mathbf{H}(J_{i,j}))$$

To prove the claim fix i. The fibre of $\mathbf{H}(\mathcal{I}_i)$ over each point of $V_{\mathcal{B},\leq_a}$ is $H^0(\mathbb{P}^2, \mathcal{I}_{I,i}(d))$ which is a subspace of $\mathcal{P}(n)$ defined by a number of linear equations that can be obtained as follows: fix a projective reference $(X : Y : Z)$, let $F = \sum_{i,j} A_{i,j}X^iY^jZ^{d-j}$ be the expression of a generic polynomial of $\mathcal{P}(d)$. We can view $A_{i,j}$ as the coordinate functions of the space $\mathcal{P}(d)$. Consider (u, v) the coordinates of a chart of X_{i-1} adapted to σ_{i-1} and centered in x_i, let $\sum_{i,j} a_{i,j}u^iv^j$ be the expression of F respect to the coordinates (u, v). Each $a_{i,j}$ is a linear function on the $A_{i,j}$'s. Because of the way of defining the ideals $\mathcal{I}_{I,i}$, the equations that define $H^0(\mathbb{P}^2, \mathcal{I}_{I,i}(d))$ as a subspace of $\mathcal{P}(d)$ are a system (S) of linear equations on the $a_{i,j}$'s whose coefficients do not depend on (I, α). To get the system of linear equations respect to the $A_{i,j}$'s, we substitute in the system (S) the variables $a_{i,j}$'s by their expressions on the $A_{i,j}$'s, getting a system $(S'_{(I,\alpha)})$. To prove the claim we have to show that we can choose a coordinate system $(u_{(I,\alpha)}, v_{(I,\alpha)})$ of a chart adapted to σ_i and centered in x_i for each point $x_i \in (I, \alpha)$ of $V_{\mathcal{B},\leq_a}$, in such a way that the coefficients of linear equations of the corresponding systems $(S'_{(I,\alpha)})$ depend algebraically on (I, α). By Lemma 3.22 we know that this can be done at least Zariski locally over $V_{\mathcal{B},\leq_a}$. □

The statement about the locally closedness of $X_{\mathcal{B}}$ can be deduced also from the techniques developed by G.M. Greuel, C. Lossen and E. Shustin in [**12**].

Consider the polynomial $\mathcal{F} \in H^0(\mathcal{P}(d), \mathcal{O}_{\mathcal{P}(d)})[X, Y, Z]$ given by

$$\mathcal{F}(X : Y : Z) = \sum_{i+j\leq d} A_{i,j}X^iY^jZ^{d-i-j},$$

where $A_{i,j}$ are like in the proof of the last proposition. By definition of the $A_{i,j}$'s, if $f \in \mathcal{P}(d)$ is of the form

$$f(X, Y, Z) = \sum_{i+j\leq d} a_{i,j}X^iY^jZ^{d-i-j},$$

then $A_{i,j}(f) = a_{i,j}$.

DEFINITION 3.27. The polynomial $\mathcal{F} \in H^0(\mathcal{P}(d), \mathcal{O}_{\mathcal{P}(d)})[X, Y, Z]$ defined in the last discussion can be viewed as a family of polynomials over $\mathcal{P}(d)$, which is called *tautological family*.

THEOREM 3.28. *Let $(\mathcal{B}, \leq_a)$ be an ordered behaviour graph with $n + 1$ vertices. Then $X_{\mathcal{B},\leq_a}$ has a natural structure of algebraic variety obtained by viewing it as a locally closed subset of $\mathcal{P}(d) \times V_{\mathcal{B},\leq_a}$. With this algebraic structure on $X_{\mathcal{B},\leq_a}$ the mapping $\tau_{\mathcal{B},\leq_a}$ becomes a morphism.*

There is a tautological family of polynomials over $X_{\mathcal{B},\leq_a}$, which will be denoted by $\mathcal{F}_{\mathcal{B}}$, obtained by taking the pullback of the tautological family of polynomials over $\mathcal{P}(d)$; for each (f, I_f, α) we have that $(\mathcal{F}_{\mathcal{B}}(f,I,\alpha), \tau_{\mathcal{B},\leq_a}(f,I,\alpha)) = (f,I,\alpha)$. Let $\mathcal{C} \subset X_{\mathcal{B},\leq_a} \times \mathbb{P}^2$ be the locus of vanishing of $\mathcal{F}_{\mathcal{B}}$. Let

$$Z_n(\mathcal{B},\leq_a) \overset{\rho_n(\mathcal{B},\leq_a)}{\longrightarrow} \ldots \overset{\rho_1(\mathcal{B},\leq_a)}{\longrightarrow} Z_0(\mathcal{B},\leq_a) = X_{\mathcal{B},\leq_a} \times \mathbb{P}^2$$

be the pullback by $\tau_{\mathcal{B},\leq_a}$ of the universal family over $V_{\mathcal{B},\leq_a}$ defined in Theorem 3.14. Define $\tilde{\mathcal{C}}_i$ to be the strict transform of $\mathcal{C}$ by the composition $\rho_1(\mathcal{B},\leq_a)\circ\ldots\circ\rho_i(\mathcal{B},\leq_a)$; then the fibre of $\tilde{\mathcal{C}}_n$ over (f,I,α) is the strict transform of the curve $Z(f)$ respect to the behaviour resolution determined by (I,α). In other words, the sequence formed by the $\rho_i(\mathcal{B},\leq_a)$'s provides a simultaneous behaviour resolution of $\mathcal{C}$.

Now we study the "forgetting ordering" mapping $\rho_{\mathcal{B},\leq_a}$; by doing this we will finally be able to prove that the subsets of the partition are algebraic:

THEOREM 3.29. *Let $(\mathcal{B},\leq_a)$ an ordered behaviour graph of degree d. The incusion $X_{\mathcal{B}} \subset \mathcal{P}(d)$ is locally closed and hence $X_{\mathcal{B}}$ is an algebraic variety. Moreover the mapping*

$$\rho_{\mathcal{B},\leq_a} : X_{\mathcal{B},\leq_a} \to X_{\mathcal{B}} \subset \mathcal{P}(d)$$

is the quotienting morphism for the natural action of $Aut(\mathcal{B})$ on $X_{\mathcal{B},\leq_a}$. As this action is free we deduce that $\rho_{\mathcal{B},\leq_a}$ is a covering.

PROOF. The mapping $\rho_{\mathcal{B},\leq_a} : X_{\mathcal{B},\leq_a} \to \mathcal{P}(d)$ is simply the restriction to $X_{\mathcal{B},\leq_a}$ of the first projection of $\mathcal{P}(d) \times V_{\mathcal{B},\leq_a}$, hence it is a morphism. We deduce that $X_{\mathcal{B}}$ is a constructible subset of $\mathcal{P}(d)$. The clousure of a constructible set in the Zariski topology is the same than its clousure in the trascendental topology. Let Z be the clousure of $X_{\mathcal{B}}$ in $\mathcal{P}(d)$. The set $X_{\mathcal{B}}$ is locally closed if and only if it is a Zariski open subset in Z. To check this it is enough to show that $X_{\mathcal{B}}$ is open in Z for the strong topology.

We will need the following notation: let h be a polynomial in one variable, we say that h is of *root type* $m_1,...,m_r$ if h has exactly r roots with multiplicities $m_1,...,m_r$.

Let f be a polynomial of $X_{\mathcal{B}}$, we assume by simplicity that the curve defined by it does not meet $(1:0:0)$, let $f(1,y,z)$ the result of dehomogeneizing f respect to the variable X. The graph $\mathcal{B}$ determines the root type $m_1,...,m_r$ of $f(1,y,0)$. Clearly, if g is another polynomial in the clousure of $X_{\mathcal{B}}$, the root type of $g(1,y,0)$ should be of the form $m'_1,...,m'_s$, where $m'_i = \sum_{j\in I_i} m_j$ for a partition $I_1,...,I_s$ of the set $\{1,...,r\}$. Let g be a polynomial of the clousure of $X_{\mathcal{B}}$ that such that the root type of $g(1,y,0)$ is different to $m_1,...,m_r$. If we want to approach g by a sequence of polynomials of $X_{\mathcal{B}}$, then the sequence is such that, as we advance in it, two points at infinity tend to collide. Take small disjoint neighbourhoods $V_1,...,V_r$ around each of the points at infinity of f. Consider the open neighbourhood U of f in $\mathcal{P}$ defined by the property that any polynomial in U should have at least m_i points at infinity in V_i counted with multiplicity. Clearly any polynomial h of $Z \cap U$ is such that the root type of $h(1,y,0)$ is $m_1,...,m_r$ (as the points at infinity are confined in $V_1,,...,V_r$ they cannot collide). Order the points at infinity of f in a compatible way with the canonical order of $\mathcal{B}$. This induces an order in the neighbourhoods $\mathcal{V}_1,...,\mathcal{V}_r$, that in turn induces an order in the points at infinity of any polynomial of $U \cap Z$. Define a mapping

$$\varphi : U \cap Z \to L^r$$

that assigns to each polynomial h the r-uple defined by the points at infinity of h with the ordering defined above. The mapping φ depends analytically on the coefficients of h and hence it is an analytic morphism.

Given any polynomial $g \in U \cap Z$ we let $p_k(g)$ denote the the k-th point at infinity of g. Let $\mathcal{B}_0$ be the result of the deletion of all the vertices of $\mathcal{B}$ except the root. It has exactly r arrows, and all of them are in Λ_2. Let a_k be the arrow corresponding to $p_k(f)$. Let (t_k, t'_k, t''_k) be the weight of a_k. For any $g \in U \cap Z$ and any $k \leq r$ we define

$$(t'_k(g), t''_k(g)) := \left(\text{mult}_{p_k(g)}(g), I_{L\cap E_k}(\mathcal{C}_k(g), E_k(g))\right)$$

where $E_k(g)$ is the exceptional divisor of the blowing up of $\mathbb{P}^2$ at $p_k(g)$ and $\mathcal{C}_k(g)$ is the strict transform of the curve defined by g by the same blowing up. We claim that for each $k \leq r$ the pair $(t'_k(g), t''_k(g))$ equals to (t'_k, t''_k). It is easy to show that for any element g of $U \cap Z$ and any $k \leq r$ we have that $(t'_k(g), t''_k(g)) \geq_{lex} (t'_k, t''_k)$ and that a necessary condition for g to be in $X_{\mathcal{B}} \cap U$ is that $(t'_k(g), t''_k(g)) = (t'_k, t''_k)$. For any i, j we define the regular functions $a^k_{i,j} : U \cap Z \to \mathbb{C}$ by the equality

$$g(1, y - y(p_k(g)), z) = \sum_{i,j} a^k_{i,j}(g) y^i z^j$$

The polynomial g satisfies $(t'_k(g), t''_k(g)) = (t'_k, t''_k)$ if and only if $a^k_{t'_k - t''_k, t''_k}(g) \neq 0$. As $(t'_k(f), t''_k(f)) = (t'_k, t''_k)$ we deduce that $a^k_{t'_k - t''_k, t''_k}(f) \neq 0$ for any $k \leq r$. Now it is clear that we can shrink U enough so that for any $g \in U \cap Z$ and any $k \leq r$ we have that $(t'_k(g), t''_k(g)) = (t'_k, t''_k)$. After this shrinking, for any $g \in U \cap Z$, the graph that we obtain by deleting all the vertices but the root of the behaviour graph of g is the same than the graph that we obtain by deleting all the vertices but the root of $\mathcal{B}$. Define $V := U \cap Z$. Let's summarize what we have done:

Let f be any polynomial of $X_{\mathcal{B}}$. We have found a neighbourhood V of f in Z such that, given any $g \in V$, the graph resulting from deleting all the vertices but the root of $\mathcal{B}$ is isomorphic to the graph obtained by deleting all the vertices but the root of $\mathcal{B}_g$. Moreover, consider an admissible ordering $\leq_a$ on $\mathcal{B}$ and fix α, an isomorphism from $\mathcal{B}_f$ to $\mathcal{B}$. The order that α induces in points at infinity of f extends to the points at infinity of any polynomial in V in such a way that the mapping $\varphi : V \to L^r$ that assigns to each polynomial $g \in V$ the r-uple defined by the points at infinity of g with the ordering defined above, is an analytic morphism.

Let $\mathcal{B}_i, \leq_a$ be the ordered graph obtained from $\mathcal{B}, \leq_a$ by deletion of all the vertices except the first i of them. The mapping $\varphi : V \to L^r$ defined above can be viewed as a mapping $\phi : V \to V_{\mathcal{B}_{r+1}, \leq_a}$. It is possible to show that for any i it is possible to shrink V so that there exists an analytic morphism $\varphi : V \to V_{\mathcal{B}_i, \leq_a}$ such that the image of any element of V is a partial resolution of it and that $\varphi(f)$ consists of the first i points of I_f respect to the ordering defined by α. The proof of these statements uses the same ideas than above. We do not make the details here because it would get lengthy.

If we let i be the number of vertices of $\mathcal{B}$ then the statement above means that we can shrink V enough to ensure that there exists an analytic morphism $\varphi : V \to V_{\mathcal{B}, \leq_a}$ such that

(1) The image $\varphi(g)$ of any $g \in V$ is a partial resolution; as $\mathcal{B}$ is a behaviour graph then $\varphi(g)$ is the behaviour resolution of f.
(2) The image of $f \in V$ is (I_f, α).

Consequently $X_{\mathcal{B}}$ is open in its clousure Z, and hence it is locally closed in $\mathcal{P}(d)$ and $\tau_{\mathcal{B},\leq_a}$ is a morphism. Moreover, for any (f, I, α) in $X_{\mathcal{B},\leq_a}$ there exists a neigboirhood V of f in $X_{\mathcal{B}}$ and an analytic section of $\rho_{\mathcal{B},\leq_a}$ that takes f to (f, I, α).

The finite group $Aut(\mathcal{B})$ acts freely in $X_{\mathcal{B},\leq_a}$. Let Y denote the geometric quotient of this action and $q : X_{\mathcal{B},\leq_a} \to Y$ the quotienting morphism. There exists a morphism $\mu : Y \to X_{\mathcal{B}}$ such that $\rho_{\mathcal{B},\leq_a} = \mu \circ q$. One checks at the level of points that q is bijective. The fact that $\rho_{\mathcal{B},\leq_a}$ has analytic local sections implies that q is an isomorphism. Hence $X_{\mathcal{B},\leq_a}$ is the geometric quotient of the free action of $Aut(\mathcal{B})$ on $X_{\mathcal{B},\leq_a}$ and $\rho_{\mathcal{B},\leq_a}$ is the quotienting morphism. □

A natural and interesting problem is to study the adjacencies of the subsets of the partition. Given two graphs $\mathcal{B}$ and $\mathcal{B}'$ of degree d, we say that $\mathcal{B}'$ is adyacent to $\mathcal{B}$ if and only if $X_{\mathcal{B}'}$ meets the cloususe of $X_{\mathcal{B}}$ in $\mathcal{P}(d)$.

PROBLEM 3.30. Consider $\mathcal{B}$ of degree d. Which graphs $\mathcal{B}'$ are adyacent to $\mathcal{B}$?. More concretely, is there a combinatorial procedure to enumerate the graphs adyacent to a given one? Is our partition an stratification?.

For work related to this see [**30**] and [**31**].

3.4. The geometry of the subsets of the partition

At this moment it is not clear which behaviour graphs are such that $X_{\mathcal{B},\leq_a}$ is not empty and what are their geometric properties. We will show that $\tau_{\mathcal{B},\leq_a}$ is surjective for *any behaviour graph* (hence $X_{\mathcal{B},\leq_a}$ is not empty). Later we will use this fact to give a rather detailed description of the geometric properties of $X_{\mathcal{B},\leq_a}$ and of the mapping $\kappa_{\mathcal{B},\leq_a}$.

PROPOSITION 3.31. *If $\mathcal{B}$ is a behaviour graph then the morphism*

$$\tau_{\mathcal{B},\leq_a} : X_{\mathcal{B},\leq_a} \to V_{\mathcal{B},\leq_a}$$

is surjective.

PROOF. Assume that the proposition is true for graphs with just one arrow, consider an arbitrary behaviour graph and an ordered sequence $I := \{x_1, ..., x_r\}$ of infinitely near points in $V_{\mathcal{B},\leq_a}$. We will construct an element of $\tau^{-1}_{\mathcal{B},\leq_a}(I)$. Let $\Lambda = \{a_1, ..., a_s\}$ be the set of arrows of $\mathcal{B}$. For any $1 \leq i \leq s$ we define the graph $\mathcal{B}_i$ as follows: remove from $\mathcal{B}$ all the arrows except a_i and delete all the vertices that are minimal in the parental order and that are not in $\gamma(a_i)$. Iterate this until the only minimal vertex is in $\gamma(a_i)$. Clearly $\mathcal{B}_i$ is a behaviour graph. Each vertex of $\mathcal{B}$ has associated a point of I. Let I_i be the subset of I formed by the points that corresponds to the vertices of $\mathcal{B}_i$. Clearly I_i is an ordered set of infinitely near points and belongs to $V_{\mathcal{B}_i,\leq_a}$. As we are assuming that the proposition is true for graphs with one arrow, there exists a polynomial $f_i \in \tau^{-1}_{\mathcal{B}_i,\leq_a}(I_i)$. It is easy to check that the behaviour graph of the product $f := f_1...f_s$ is isomorphic to $\mathcal{B}$ and that, under the right choice of an isomorphism $\alpha : \mathcal{B}_f \to \mathcal{B}$, the triple (f, I_f, α) is an element of $\tau^{-1}_{\mathcal{B},\leq_a}(I)$.

We have reduced the proof to the case of behaviour graphs with only one arrow. We treat this in the series of lemmas that follows. □

LEMMA 3.32. *Let $\mathcal{B}$ be any behaviour graph with one arrow such that its parental system is contained in the parental system of the graph $\mathcal{A}_n$ for a certain n. Then $\tau_{\mathcal{B},\leq_a}$ is surjective.*

PROOF. There are three types of graphs satisfying the hypothesis:

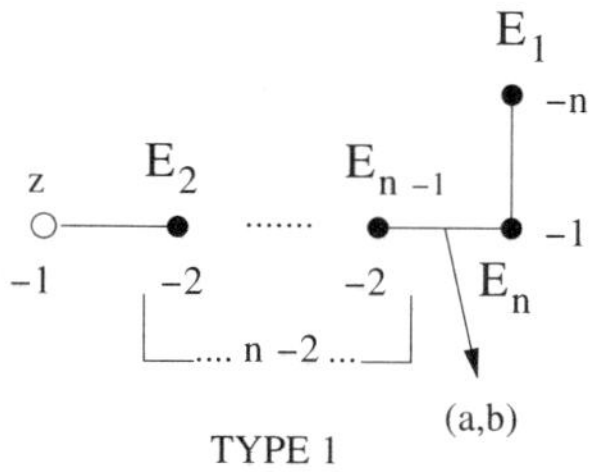

TYPE 1

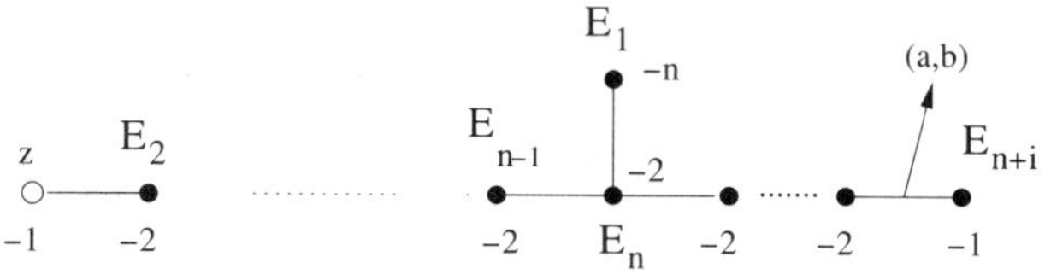

TYPE 2

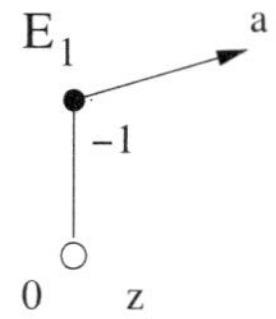

TYPE 3

If $\mathcal{B}$ is of type 1 or 3 there is only one free vertex, namely the vertex representing E_1, hence in these cases $V_{\mathcal{B},\leq_a}$ is canonically isomorphic to the line at infinity L. If $\mathcal{B}$ is of type 1 then the polynomial $x^{(n-1)a+nb} + y^{a+b}$ is an element of the fibre of $\tau_{\mathcal{B},\leq_a}$ over $(0:1:0) \in L$. If $\mathcal{B}$ is of type 3 then the polynomial y^a is an element of the fibre of $\tau_{\mathcal{B},\leq_a}$ over $(0:1:0) \in L$. In both cases to get elements of the fibres of $\tau_{\mathcal{B},\leq_a}$ over other points of $p \in L$ we compose the given element of the fibre over $(0:1:0)$ by a linear transformation that takes p into $(0:1:0)$.

Suppose that $\mathcal{B}$ is of type 2. Let I be an element of $V_{\mathcal{B},\leq_a}$. Let $\pi_0 := X_0 \to \mathbb{P}^2$ be the unique admissible model for I. Choose a blowing up process $\pi_1 : X \to X_0$ such that $\pi := \pi_0 \circ \pi_1$ belongs to A_n. Let $\phi := t_n(\pi)$, and consider its factorization in blowing ups and blowing downs $\phi = \pi' \circ \pi^{-1}$. Let $\mathcal{B}_0$ be a behaviour of type 1 with $n-i+2$ vertices and such that the weight of the arrow is (b,a). Let q be the first blowing up center of $\pi' : X \to \mathbb{P}^2$. Consider a polynomial f that is in the fibre over q of $\tau_{\mathcal{B}_0,\leq_a}$. Then the polynomial $f \circ \phi$ is an element of the fibre of $\tau_{\mathcal{B},\leq_a}$ over I. □

Next we remove the undesired hypothesis:

LEMMA 3.33. *Let $\mathcal{B}$ be any behaviour graphs with one arrow. Then $\tau_{\mathcal{B},\leq_a}$ is surjective.*

PROOF. Let $I \in V_{\mathcal{B},\leq_a}$, we define I_0 to be the maximal subset of I such that its associated blowing up process π_0 is in $A_{n_1,\ldots,n_r}$ for some $n_1, \ldots, n_r$; let $I_1 := I \setminus I_0$. Consider $\phi := t_{n_1,\ldots,n_r}(\pi_0)$ and view it as a birational map from $\mathbb{P}^2$ to $\mathbb{P}^2$. The set I_1 is a saturated set of infinitely near points for the target $\mathbb{P}^2$. Define the graph $\mathcal{B}_1$ obtained by deleting from $\mathcal{B}$ all the vertices of I_0 except the last one. The set I_1 defines an element of $V_{\mathcal{B}_1,\leq_a}$. As $\mathcal{B}_1$ has only one floor there is a polynomial f

in the fibre of $\tau_{\mathcal{B},\leq_a}$ over I_1. Then the polynomial $f \circ \phi$ is an element of the fibre of $\tau_{\mathcal{B},\leq_a}$ over I. □

REMARK 3.34. With Proposition 3.31 we have shown that every behaviour graph is realizable by a polynomial

In the rest of the section we will use Proposition 3.31 as a tool to study the geometry of the morphism $\tau_{\mathcal{B},\leq_a}$ for a fixed behaviour graph $\mathcal{B}$.

LEMMA 3.35. *Let $\mathcal{B}$ be a behaviour graph and $\leq_a$ be an admissible ordering on it. Let $d = \deg(\mathcal{B})$. Consider $(I, \alpha) \in V_{\mathcal{B},\leq_a}$. Then the fibre $\tau_{\mathcal{B},\leq_a}^{-1}(I, \alpha)$ is equal to*

$$H^0(\mathbb{P}^2, \mathcal{I}_I(d)) \setminus \cup_{i,j} H^0(\mathbb{P}^2, \mathcal{J}_{I,i,j}(d)) \tag{3.6}$$

where i, j run over all the values for which $\mathcal{J}_{I,i,j}$ has been defined. Each of the $H^0(\mathbb{P}^2, \mathcal{J}_{I,i,j}(d))$ is a hyperplane in $H^0(\mathbb{P}^2, \mathcal{I}_I(d))$. Hence $\tau_{\mathcal{B},\leq_a}^{-1}(I, \alpha)$ is the complement of a hyperplane arrangement in a vector space.

PROOF. The lemma follows from Lemma 3.25 and the fact that $\tau_{\mathcal{B},\leq_a}$ is surjective: as the fibre $H^0(\mathbb{P}^2, \mathcal{I}_I(d)) \setminus \cup_{i,j} H^0(\mathbb{P}^2, \mathcal{J}_{I,i,j}(d))$ is non empty the subspace $H^0(\mathbb{P}^2, \mathcal{J}_{I,i,j}(d))$, which is linear of codimension at most 1, has to be a hyperplane. □

PROPOSITION 3.36. *The morphism $\tau_{\mathcal{B},\leq_a}$ is equidimensional.*

PROOF. We need to show that, $h^0(\mathbb{P}^2, \mathcal{I}_I(d))$ does not depend of I when we vary (I, α) in $V_{\mathcal{B},\leq_a}$. For any $(I, \alpha) \in V_{\mathcal{B},\leq_a}$ we consider the exact sequence of sheaves

$$0 \to \mathcal{I}_I \to \mathcal{O}_{\mathbb{P}^2} \to \mathcal{O}_{\mathbb{P}^2}/\mathcal{I}_I \to 0$$

Noticing that $H^1(\mathbb{P}^2, \mathcal{O}_{\mathbb{P}^2}(d)) = 0$, and that $\mathcal{O}_{\mathbb{P}^2}/\mathcal{I}_I$ is a direct sum of skyscraper sheaves we get the following exact sequence

$$0 \to H^0(\mathbb{P}^2, \mathcal{I}_I(d)) \to H^0(\mathbb{P}^2, \mathcal{O}_{\mathbb{P}^2}(d)) \to \mathcal{O}_{\mathbb{P}^2}/\mathcal{I}_I \to H^1(\mathbb{P}^2, \mathcal{I}_I(d)) \to 0 \tag{3.7}$$

If we prove that the map $\mu : H^0(\mathbb{P}^2, \mathcal{O}_{\mathbb{P}^2}(d)) \to \mathcal{O}_{\mathbb{P}^2}/\mathcal{I}_I$ is surjective then

$$h^0(\mathbb{P}^2, \mathcal{I}_I(d)) = h^0(\mathbb{P}^2, \mathcal{O}_{\mathbb{P}^2}(d)) - \mathrm{length}(\mathcal{O}_{\mathbb{P}^2}/\mathcal{I}_I)$$

and hence showing that $\mathrm{length}(\mathcal{O}_{\mathbb{P}^2}/\mathcal{I}_I)$ is independent of I would finish the proof. The needed asertions will be the object of the next lemma. □

LEMMA 3.37. *The homomorphism $\mu : H^0(\mathbb{P}^2, \mathcal{O}_{\mathbb{P}^2}(d)) \to \mathcal{O}_{\mathbb{P}^2}/\mathcal{I}_I$ is surjective and* $\mathrm{length}(\mathcal{O}_{\mathbb{P}^2}/\mathcal{I}_I)$ *is independent of I.*

PROOF. To show this we will construct a set $A = \{f_1, ..., f_l\} \subset H^0(\mathbb{P}^2, \mathcal{O}_{\mathbb{P}^2}(d))$ that maps via μ to a basis of $\mathcal{O}_{\mathbb{P}^2}/\mathcal{I}_I$. The way of doing it is to construct a filtration of intermediate ideals

$$\mathcal{O}_{\mathbb{P}^2} = \mathcal{I}'_0 \supset \mathcal{I}'_1 \supset ... \supset \mathcal{I}'_l = \mathcal{I}_I$$

such that for any $I \leq l$ we have that $\mathrm{length}(\mathcal{I}'_{i-1}/\mathcal{I}'_i) = 1$ and to construct, for each $1 \leq i \leq l$ a polynomial

$$f_i \in H^0(\mathbb{P}^2, \mathcal{I}'_{i-1}(d)) \setminus H^0(\mathbb{P}^2, \mathcal{I}'_i(d)).$$

The filtration of intermediate ideals is constructed as follows: let

$$l_i := \mathrm{length}(\cap_{j=1}^{i} \mathcal{I}_{I,j} / \cap_{j=1}^{i-1} \mathcal{I}_{I,j}).$$

Let $n+1$ be the number of vertices of $\mathcal{B}$. For any $1 \leq i \leq n$ define

$$\mathcal{I}'_{\sum_{j=1}^{i} l_j} := \cap_{j=1}^{i} \mathcal{I}_{I,j} \quad \mathcal{I}_0 := \mathcal{O}_{\mathbb{P}^2}.$$

To construct the ideals between $\mathcal{I}'_{\sum_{j=1}^{i-1} l_j}$ and $\mathcal{I}'_{\sum_{j=1}^{i} l_j}$ we will use the notations of construction 3.23. Let $\mathcal{B}_{i-1}$ be the partial resolution graph obtained by deleting all the vertices of $\mathcal{B}$ smaller or equal than v_i. Let a_i the arrow that corresponds to v_i.

Case 1. Suppose that $a_i \in \Lambda_2$. Let v_j be the attaching vertex of a_i. Let (t, t', t'') be the weight of a_i. Consider (x, y), the coordinates of a chart of X_{i-1} adapted to σ_{i-1} and centered at x_i (in such a chart E_j is defined by $y = 0$). Let f be any germ of function regular at $\sigma_{i-1}(x_i)$, let $\sigma_{i-1}^* f(x, y) = \sum a_{i,j} x^i y^j$ be the expression of $\sigma_{i-1}^* f$ respect to (x, y). by the definition of the ideal sheaf $\mathcal{I}'_{\sum_{j=1}^{i-1} l_j}$ it is easy to conclude that $y^{\mathrm{mtot}(v_j)}$ divides $\sigma_{i-1}^* f(x, y)$ for any germ $f \in \mathcal{I}'_{\sum_{j=1}^{i-1} l_j}$. Moreover, if f is generic in $\mathcal{I}'_{\sum_{j=1}^{i-1} l_j}$ then $\sigma_{i-1}^* f(x, y) = y^{\mathrm{mtot}(v_j)} u$ where $u(x_i) \neq 0$.

We will define intermediate ideals $\mathcal{I}'_{\sum_{j=1}^{i-1} l_j + s}$ for certain $s \geq 0$ in the following way: define $\mathcal{I}'_{\sum_{j=1}^{i-1} l_j + s}$ to be the sheaf of functions f regular at $\sigma_{i-1}(x_i)$ whose coefficients respect (x, y) satisfy the following linear conditions:

- If $s \leq t$ then the conditions are

$$a_{i,\mathrm{mtot}(v_j)} = 0 \quad \text{if} \quad i < s.$$

- If $t + \frac{(k-1)k}{2} < s \leq t + \frac{k(k+1)}{2}$ for $k < t'$ the conditions are

$$a_{i,\mathrm{mtot}(v_j)+j} = 0 \quad \text{if} \quad i + j < k,$$

$$a_{i,\mathrm{mtot}(v_j)+j} = 0 \quad \text{if} \quad i + j = k \quad \text{and} \quad j \leq s - (t + \frac{(k-1)k}{2}).$$

- If $t'' \leq 1$ then the ideal $\mathcal{I}'_{(\sum_{j=1}^{i-1} l_j)+t+\frac{(t'-1)t'}{2}}$ (which is the last ideal defined above) equals $\mathcal{I}'_{\sum_{j=1}^{i-1} l_j} \cap \mathcal{I}_{I,i}$ which is the same than $\mathcal{I}'_{\sum_{j=1}^{i} l_j}$. Hence we do not have to define more intermediate ideals. Assume that $t'' > 1$. If $t + \frac{(t'-1)t'}{2} < s \leq t + \frac{(t'-1)t'}{2} + t'' - 1$ then the conditions are

$$a_{i,\mathrm{mtot}(v_j)+j} = 0 \quad \text{if} \quad i + j < t',$$

$$a_{i,\mathrm{mtot}(v_j)+j} = 0 \quad \text{if} \quad i + j = t' \quad \text{and} \quad j \leq s - (t + \frac{(t'-1)t'}{2}).$$

The ideal

$$\mathcal{I}'_{(\sum_{j=1}^{i-1} l_j)+t+\frac{(t'-1)t'}{2}+\max(0,t''-1)}$$

(which is the last ideal defined above) equals $\mathcal{I}'_{\sum_{j=1}^{i-1} l_j} \cap \mathcal{I}_{I,i}$ which is the same than $\mathcal{I}'_{\sum_{j=1}^{i} l_j}$.

Case 2. Suppose that $a_i \in \Lambda_4$. Let $\gamma(a_i) = \{v_j, v_k\}$, with $j < k$. Let $(t_1, t_2, t_3, t_4, t_5)$ be the weight of a_i. Consider (x, y), the coordinates of a chart of X_{i-1} adapted to σ_{i-1} and centered at x_i (in such a chart E_k is defined by $y = 0$ and E_j by $x = 0$). Let f be any germ of function regular at $\sigma_{i-1}(x_i)$, let $\sigma_{i-1}^* f(x, y) = \sum a_{i,j} x^i y^j$ be the expression of $\sigma_{i-1}^* f$ respect to (x, y). By the definition of the ideal sheaf $\mathcal{I}'_{\sum_{j=1}^{i-1} l_j}$ it is easy to conclude that if f is any generic germ at $\sigma_{i-1}(x_i)$ in the ideal $\mathcal{I}'_{\sum_{j=1}^{i-1} l_j}$ then the expression of $\sigma_{i-1}^* f$ respect to (x, y)

is of the form $x^{\mathrm{mtot}(v_j)}y^{\mathrm{mtot}(v_k)}h(x,y)$, where $h(x,y)$ defines a germ of curve $V(h)$ at x_i that satisfies

$$\mathrm{mult}(h) \leq 2 \quad I_{x_i}(E_j, V(h)) = t_1 \quad I_{x_i}(E_k, V(h)) = t_2$$

We will define intermediate ideals $\mathcal{I}'_{\sum_{j=1}^{i-1} l_j+s}$ for certain $s \geq 0$ in the following way: define $\mathcal{I}'_{\sum_{j=1}^{i-1} l_j+s}$ to be the sheaf of functions f regular at $\sigma_{i-1}(x_i)$ whose coefficients respect (x,y) satisfy the following linear conditions:

- If either t_1 or t_2 equal 1 then any element of $\mathcal{I}'_{\sum_{j=1}^{i-1} l_j}$ belongs to $\mathcal{I}_{I,i}$. Then $\mathcal{I}'_{\sum_{j=1}{i-1} l_j} = \mathcal{I}'_{\sum_{j=1}^{i} l_j}$ and we do not have to define intermediate ideals at all. Assume that $t_1 > 1$ and that $t_2 > 1$. If $\frac{(k-2)(k-1)}{2} < s \frac{(k-1)k}{2}$ for $k < t_3$ then the conditions are

$$a_{\mathrm{mtot}(v_j)+i,\mathrm{mtot}(v_k)+j} = 0 \quad \text{if} \quad i+j < k+1,$$

$$a_{\mathrm{mtot}(v_j)+i,\mathrm{mtot}(v_k)+j} = 0 \quad \text{if} \quad i+j = k+1 \quad \text{and} \quad j \leq s - \frac{(k-1)k}{2}.$$

- If $t_4 \leq 1$ go to the next item. If $t_4 > 1$ and

$$\frac{(t_3-2)(t_3-1)}{2} < s \leq \frac{(t_3-2)(t_3-1)}{2} + t_4 - 1$$

  then the conditions are

$$a_{\mathrm{mtot}(v_j)+i,\mathrm{mtot}(v_k)+j} = 0 \quad \text{if} \quad i+j < t_3,$$

$$a_{\mathrm{mtot}(v_j)+i,\mathrm{mtot}(v_k)+j} = 0 \quad \text{if} \quad i+j = t_3 \quad \text{and} \quad i \leq s - \frac{(t_3-2)(t_3-1)}{2}.$$

- If $t_5 \leq 1$ then the ideal $\mathcal{I}'_{(\sum_{j=1}^{i-1} l_j)+\frac{(t_3-2)(t_3-1)}{2}+\max(0,t_4-1)}$ (which is the last ideal defined above) equals $\mathcal{I}'_{\sum_{j=1}^{i-1} l_j} \cap \mathcal{I}_{I,i}$ which is the same than $\mathcal{I}'_{\sum_{j=1}^{i} l_j}$. Hence we do not have to define more intermediate ideals. If $t_5 > 1$ and

$$\frac{(t_3-2)(t_3-1)}{2} + \max(0, t_4-1) < s$$

$$s \leq \frac{(t_3-2)(t_3-1)}{2} + \max(0, t_4-1) + t_5 - 1$$

  then the conditions are

$$a_{\mathrm{mtot}(v_j)+i,\mathrm{mtot}(v_k)+j} = 0 \quad \text{if} \quad i+j < t_3,$$

$$a_{\mathrm{mtot}(v_j)+i,\mathrm{mtot}(v_k)+j} = 0 \quad \text{if} \quad i+j = t_3 \quad \text{and} \quad i \leq t_4 - 1,$$

$$a_{\mathrm{mtot}(v_j)+i,\mathrm{mtot}(v_k)+j} = 0 \quad \text{if} \quad i+j = t_3 \quad \text{and}$$

$$j \leq s - \Big(\frac{(t_3-2)(t_3-1)}{2} + \max(0, t_4-1)\Big).$$

  The ideal $\mathcal{I}'_{(\sum_{j=1}^{i-1} l_j)+\frac{(t_3-2)(t_3-1)}{2}+\max(0,t_4-1)+\max(0,t_5-1)}$ (which is the last ideal defined above) equals $\mathcal{I}'_{\sum_{j=1}^{i-1} l_j} \cap \mathcal{I}_{I,i}$ which is the same than $\mathcal{I}'_{\sum_{j=1}^{i} l_j}$. Hence we do not have to define more intermediate ideals.

Notice that, for any i, we have constructed each $\mathcal{I}'_i$ by adding a linear condition to $\mathcal{I}'_{i-1}$, this implies that length$(\mathcal{I}'_{i-1}/\mathcal{I}'_i) = 1$, and hence the $\mathcal{I}'_i$'s form the desired sequence of ideals.

As a by-product we have shown that if v_i is free and (t, t', t'') is the weight of the arrow that corresponds to v_i in $\mathcal{B}_{i-1}$ then

$$\text{length}(\cap_{j=1}^{i-1}\mathcal{I}_{I,i}/\cap_{j=1}^{i}\mathcal{I}_{I,i}) = t + \frac{(t'-1)t'}{2} + \max(0, t''-1), \tag{3.8}$$

and if v_i is fixed and $(t_1, t_2, t_3, t_4, t_5)$ is the weight of the arrow that corresponds to v_i in $\mathcal{B}_{i-1}$ then

$$\text{length}(\cap_{j=1}^{i-1}\mathcal{I}_{I,i}/\cap_{j=1}^{i}\mathcal{I}_{I,i}) = \frac{(t_3-2)(t_3-1)}{2} + \max(0, t_4-1) + \max(0, t_5-1). \tag{3.9}$$

Notice that the last expression is valid also when t_1 or t_2 are equal to 1: in this case t_3 has to be also equal to 1 because $\mathcal{B}$ is a behaviour graph, and the expresion 3.9 vanish. On the other hand we have seen that if t_1 or t_2 equals one then $\cap_{j=1}^{i-1}\mathcal{I}_{I,i} = \cap_{j=1}^{i}\mathcal{I}_{I,i}$. The last formulae imply that length$(\mathcal{O}_{\mathbb{P}^2}/\mathcal{I}_I)$ does not depend on I.

Finding, for each k, an f_k in $H^0(\mathbb{P}^2, \mathcal{I}'_k(d)) \setminus H^0(\mathbb{P}^2, \mathcal{I}'_{k+1}(d))$, would finish the proof of the lemma. If $k = l_1 + ... + l_r$ then $\mathcal{I}'_k$ is the smallest ideal of the filtration, and hence any polynomial of $\tau^{-1}_{\mathcal{B},\leq_a}(I, \alpha)$ would serve as f_k. Assume that $k < l_1 + ... + l_r$. Consider the unique i such that $\sum_{j=1}^{i-1} l_j \leq k < \sum_{j=1}^{i} l_j$. Take $s := k - \sum_{j=1}^{i-1} l_j$. Let a_i the arrow that is associated with v_i in $\mathcal{B}_{i-1}$. We study the case in which $\gamma(a_i) = v_j$. Let (t, t', t'') be the weight of a_i. We construct a graph $\mathcal{G}_k$ by introducing the following slight modification on $\mathcal{B}_{i-1}$:

- If $s = 0$ then delete a_i and add a new arrow b with the same weight than a_i.
- If $s = 1$ then we change the weight of a_i by $(1, 1, 0)$ and we attach a new arrow b to v_j with weight (u, u', u'') where $(u, u', u'') = (t-1, 1, 1)$ if $t > 2$ or $(u, u', u'') = (1, 1, 0)$ if $t = 2$ (note that $t \neq 1$ because otherwise k would be equal to $\sum_{j=1}^{i} l_j$).
- If $1 < s < t$ then we change the weight of a_i by $(s, 1, 1)$ and we attach a new arrow b to v_j with weight (v, v', v'') where $(v, v', v'') = (t - s, 1, 1)$ if $t - s > 1$ and $(v, v', v'') = (1, 1, 0)$ if $t - s = 1$.
- If $t + \frac{(k-1)k}{2} \leq s < t + \frac{k(k+1)}{2}$ for $1 < k < t'$ then we change the weight of a_i by $(t, k, 1 + s - (t + \frac{(k-1)k}{2}))$.
- If $t + \frac{(t'-1)t'}{2} \leq s < t + \frac{(t'-1)t'}{2} + max(0, t''-1)$ then we change the weight of a_i by $(t, k, 1 + s - (t + \frac{(t'-1)t'}{2}))$.

The graph $\mathcal{G}_k$ is a decorated parental graph. An easy exercise in combinatorics shows that there exists a behaviour graph $\mathcal{H}_k$ such that $\mathcal{G}_k$ can be obtained from $\mathcal{H}_k$ by deletion of vertices. The set $\{v_1, ..., v_{i-1}\}$ is the set of vertices of $\mathcal{G}_k$. Let $\{v_1, ..., v_{i-1}, w_i, ..., w_{r'}\}$ be the set of vertices of $\mathcal{H}_k$. Consider an admissible ordering $\leq_a$ in $\mathcal{H}_k$. Remember that $I = \{x_1, .., x_r\}$ is the element of $V_{\mathcal{B},\leq_a}$ with which we are working. We can choose an element $I' = \{y_1, ..., y_{r'}\} \in V_{\mathcal{H}_k,\leq_a}$ such that

- For each $i \leq i - 1$ we have that $y_i = x_i$.
- In case that the arrow b were added in the construction of $\mathcal{G}_k$, the infinitely near point of I' corresponding to it does not belong to $\{x_i, ..., x_r\}$.

- If a_i exists in $\mathcal{G}_k$ the infinitely near point of I' corresponding to the vertex of $\mathcal{H}_k$ whose deletion gives a_i is x_i.

By Proposition 3.31 we know that there exist a triple (f_k, J, β) such that

$$(f_k, J, \beta) \in \tau^{-1}_{\mathcal{H}_e, \leq_a}(\{y_1, ..., y_{r'}\})$$

Because of the way of constructing $\mathcal{H}_k$ and $\{y_1, ..., y_{r'}\}$ we know that

$$f_k \in H^0(\mathbb{P}^2, \mathcal{I}'_k(d)) \setminus H^0(\mathbb{P}^2, \mathcal{I}'_{k+1}(d))$$

as required.

The case in which $\gamma(a_i)$ is a set of two vertices can be treated using analogous techniques and we leave it to the reader. □

A corollary of the last lemma and exact sequence 3.7 is

COROLLARY 3.38. *Given any behaviour graph $\mathcal{B}$ whose degree is d and a point $(I, \alpha) \in V_{\mathcal{B}, \leq_a}$ we have*

$$H^1(\mathbb{P}^2, \mathcal{I}_I(d)) = 0$$

Vanishing theorems similar to the one above have been quite important in the work of Greuel, Lossen and Shustin for proving theorems on non emptyness, irreducibility and smoothness of varieties parametrizing curves of a fixed degree with singularities of a prescribed type, (see [**11**],[**12**],[**13**]). The vanishing theorem above is new, and I could not derive it using the techniques from the authors cited above. If instead of behaviour resolutions we would use embedded resolution of the singularities at infinity, I do not know how to derive the corresponding vanishing theorem; here is one of the places where the advantage of using behaviour resolutions becomes clear.

Let $\mathcal{B}$ be a behaviour graph, $\leq_a$ an admissible ordering and $(I, \alpha) \in V_{\mathcal{B}, \leq_a}$. In Lemma 3.35 we have shown that $\tau^{-1}_{\mathcal{B}, \leq_a}(I, \alpha)$ is the complement in $H^0(\mathbb{P}^2, \mathcal{I}_S(d))$ of the hyperplane arrangement $\cup_{i,j} H^0(\mathbb{P}^2, \mathcal{J}_{I,i,j}(d))$. We will reinterprete the meaning of the hyperplanes $H^0(\mathbb{P}^2, \mathcal{J}_{I,i,j}(d))$ and show that some of them may coincide.

DEFINITION 3.39. Let $(I, \alpha) \in V_{\mathcal{B}, \leq_a}$. Let $\sigma_i : X_i \to \mathbb{P}^2$ be the composition of the blowing ups centered in the i first vertices of I. Take $z \in E_i$, $g \in H^0(\mathbb{P}^2, \mathcal{I}_I(d))$, and $(V, (x, y))$ an affine chart adapted to σ_i of X_i centered at z. We define the *virtual strict transform* of g by σ_i at z respect $(V, (x, y))$ to be the polynomial

$$\frac{1}{y^{\mathrm{mtot}(v_i)}} \sigma_i^* g$$

if E_i is the only component of $\sigma_i^* L$ that z meets, and the polynomial

$$\frac{1}{x^{\mathrm{mtot}(v_j)} y^{\mathrm{mtot}(v_i)}} \sigma_i^* g$$

if $z = E_i \cap E_j$. The zero locus of the virtual strict transform of f by σ_i will be denoted by $\tilde{C}_i$.

REMARK 3.40. The virtual strict transform by σ_i of a polynomial f such that $(f, I_f, \alpha) \in \tau^{-1}_{\mathcal{B}, \leq_a}(I, \alpha)$ for certain α coincides with its strict transform by σ_i.

Consider v_i an vertex of $\mathcal{B}$ that is not a free central vertex. There is a fixed set of points $\{y_1, ..., y_s\} \subset I$ and a fixed sequence of numbers $(d_1, ..., d_s)$ such that, if (f, I, α) is any element of $\tau^{-1}_{\mathcal{B}, \leq_a}(I, \alpha)$, then $E_i \cap \tilde{C}_i = \{y_1, ..., y_s\}$ and

$I_{y_s}(E_i, \tilde{C}_i) = d_i$. A consequence of this is the following: let $(V, (x, y))$ be an affine chart of X_i such that $E_i \cap V$ is determined by $x = 0$. Consider (f_1, I, α) and (f_2, I, α), elements of $\tau^{-1}_{\mathcal{B},\leq_a}(I, \alpha)$, let $g_1(x, y)$ and $g_2(x, y)$ be the equation of the strict transform of f_1 and f_2 by σ_i. *Then $g_1(0, y)$ should be proportional (by a non zero constant) to $g_2(0, y)$*. From here one deduces easily the following

LEMMA 3.41. *Let v_i be a vertex of $\mathcal{B}$ that is not a free central vertex. Given any $f_1, f_2 \in H^0(\mathbb{P}^2, \mathcal{I}_I(d))$, then, if g_1, g_2 are their virtual strict transforms by σ_i respect to an affine chart $(V, (x, y))$ such that x defines E_i, then $g_1(0, y)$ and $g_2(0, y)$ are proportional.*

With the notations introduced above we have that $\mathrm{mtot}(v_i) = d_1 + ... + d_s$. If (f, I_f, α) is any element of $\tau^{-1}_{\mathcal{B},\leq_a}(I, \alpha)$ then the multiplicity of the strict transform $\tilde{C}_{i-1}$ at x_i is exactly $\mathrm{mtot}(v_i)$. Take any $f \in H^0(\mathbb{P}^2, \mathcal{I}_I(d))$, let g be the virtual strict transform of f by σ_i respect to an affine chart as above. We have that $g(0, y) \equiv 0$ if and only if $\mathrm{mult}_{x_i}(\tilde{C}_{i-1}) > \mathrm{mtot}(v_i)$. by Lemma 3.41 we know that the condition $g(0, y) \equiv 0$ defines a hyperplane in $H^0(\mathbb{P}^2, \mathcal{I}_I(d))$. And hence the condition $\mathrm{mult}_{x_i}(\tilde{C}_{i-1}) > \mathrm{mtot}(v_i)$ also. We denote this hyperplane by H_{v_i}.

Let v_i be a free central vertex such that there is no arrow of $\mathcal{B}$ attached to it. Let $p(v_i) = v_j$. Let $f \in H^0(\mathbb{P}^2, \mathcal{I}_I(d))$. Consider $(V, (x, y))$, an affine chart of X_i adapted to σ_i and centered at $E_i \cap E_j$. Let g be the virtual strict transform of f by σ_i respect to $(V, (x, y))$. An easy argument shows that $g(0, y)$ is of the form $g(0, y) = Ay^{\mathrm{mtot}(v_i)}$. We deduce that $\mathrm{mult}_{x_i}(\tilde{C}_{i-1}) > \mathrm{mtot}(v_i)$ if and only if $A = 0$. Hence that the condition $\mathrm{mult}_{x_i}(\tilde{C}_{i-1}) > \mathrm{mtot}(v_i)$ defines a hyperplane in $H^0(\mathbb{P}^2, \mathcal{I}_I(d))$. We denote this hyperplane by H_{v_i}.

Let v_i be a free central vertex such that there is an arrow of $\mathcal{B}$ attached to it. Let $p(v_i) = v_j$. Let (f, I_f, α) be any element of $\tau^{-1}_{\mathcal{B},\leq_a}(I)$. The graph $\mathcal{B}$ prescribes $I_{E_i \cap E_j}(\tilde{C}_i, E_i)$, let this number be t. One checks easily than the condition $I_{E_i \cap E_j}(\tilde{C}_i, E_i) > t$ defines a hyperplane in $H^0(\mathbb{P}^2, \mathcal{I}_I(d))$. We denote this hyperplane by H_{v_i}

DEFINITION 3.42. Let v be any vertex of $\mathcal{B}$. If v is the root we define H_v to be the hyperplane of $H^0(\mathbb{P}^2, \mathcal{I}_I(d))$ that consists of the polynomials that contain the line at infinity as a component. If v is not the root we define H_v as in the last discussion.

PROPOSITION 3.43. *Let $\mathcal{B}$ be a behaviour graph and $(I, \alpha) \in V_{\mathcal{B},\leq_a}$. Then for any i, j there exist a vertex $v \in \mathcal{B}$ such that*

$$H^0(\mathbb{P}^2, \mathcal{J}_{I,i,j}(d)) = H_v.$$

PROOF. Given any $(f, I_f, \alpha) \in \tau^{-1}_{\mathcal{B},\leq_a}(I, \alpha)$ the graph $\mathcal{B}$ prescribes the numbers

$$I_{E_r \cap E_s}(\tilde{C}_k, E_r) \quad I_{E_r \cap E_s}(\tilde{C}_k, E_s)$$

for any r, s, k for which the expressions above make sense. It also prescribes $I_{x_i}(\tilde{C}_k, E_r)$ for any $x_i \in I$ that has a representative in E_r. We denote these numbers by $i_{E_r \cap E_s}(\tilde{C}_k, E_r)$, $i_{E_r \cap E_s}(\tilde{C}_k, E_s)$ and $i_{x_i}(\tilde{C}_k, E_r)$ respectively.

Consider any $g \in H^0(\mathbb{P}^2, \mathcal{J}_{I,i,j}(d))$. From the definition of $\mathcal{J}_{I,i,j}$ (Definition 3.24) we deduce that

- If $p(v_i) = v_k$ then at least one of the following strict inequalities holds:

$$I_{E_k \cap E_i}(\tilde{C}_i, E_i) > i_{E_k \cap E_i}(\tilde{C}_i, E_i) \tag{3.10}$$

(3.11) $$I_{E_k\cap E_i}(\tilde{C}_i, E_k) > i_{E_k\cap E_i}(\tilde{C}_i, E_k)$$

- If $p(v_i) = \{v_k, v_l\}$ then at least one of the following strict inequalities holds:

$$I_{E_k\cap E_i}(\tilde{C}_i, E_k) > i_{E_k\cap E_i}(\tilde{C}_i, E_k) \quad I_{E_l\cap E_i}(\tilde{C}_i, E_l) > i_{E_l\cap E_i}(\tilde{C}_i, E_l)$$
$$I_{E_k\cap E_i}(\tilde{C}_i, E_i) > i_{E_k\cap E_k}(\tilde{C}_i, E_i) \quad I_{E_l\cap E_i}(\tilde{C}_i, E_i) > i_{E_l\cap E_i}(\tilde{C}_i, E_i)$$

We study the case in which $p(v_i) = v_k$. Suppose that the inequality 3.10 holds. If v_i is free central then, by the definition of H_{v_i}, we have that $H^0(\mathbb{P}^2, \mathcal{J}_{I,k,j}(d)) \subset H_{v_k}$. As both are hyperplanes they coincide. If v_i is not a free central vertex we proceed as follows: let $(V, (x, y))$ be an affine chart of X_i such that E_i and E_k are defined by $x = 0$ and $y = 0$ respectively; consider any generic element $f \in H^0(\mathbb{P}^2, \mathcal{I}_I(d))$, if we denote by f_k the strict transform of f by σ_k with respect to the chart $(V, (x, y))$, then

$$f_k(0, y) = Ay^{i_{E_k\cap E_i}(\tilde{C}_i, E_i)} + \quad \text{higher order terms}$$

If g_k is the virtual strict transform of g with respect to the same chart the inequality 3.10 implies that $g(0, y) \in (y)^{i_{E_k\cap E_i}(\tilde{C}_k, E_k)+1}$. Using Lemma 3.41 we deduce that $g(0, y) \equiv 0$ and hence that $\text{mult}_{x_i}(\tilde{C}_{i-1})$ is strictly bigger than the value prescribed by $\mathcal{B}$. This implies that $H^0(\mathbb{P}^2, \mathcal{J}_{I,k,j}(d)) = H_{v_k}$.

Now we suppose that the inequality 3.11 holds. This implies that $I_{x_i}(\tilde{C}_k, E_k) > i_{x_i}(\tilde{C}_k, E_k)$. An argument like before shows that, if v_k is not the root, $\text{mult}_{x_k}(\tilde{C}_{k-1})$ is strictly bigger than the value prescribed by $\mathcal{B}$, and that, if v_k is the root, g should contain the line at infinity as a component. In any case we show that $H^0(\mathbb{P}^2, \mathcal{J}_{I,k,j}(d)) = H_{v_k}$.

The case in which $p(v_k) = \{v_i, v_j\}$ is treated similarly and we leave it to the reader. □

Definition 3.44. Let $v \in \mathcal{B}$ and $v' \in p(v)$. We say that v *depends strictly* on v' if and only if there is neither an arrow a of $\mathcal{B}$ such that $\gamma(a) = \{v, v'\}$ nor a vertex v'' such that $p(v'') = \{v, v'\}$. Let v and v' arbitrary vertices of $\mathcal{B}$. We say that v and v' *are dependent* if and only if there is a sequence $v = v_1, ..., v_k = v'$ such that for any $1 \le i \le n - 1$ either the vertex v_i depends strictly of v_{i+1} or the vertex v_{i+1} depends strictly on v_i.

Remark 3.45. Two vertices of $\mathcal{B}$ are dependent if and only if they are in the same conected component of the result of deleting from $\mathcal{B}$ all the edges that have an arrow attached.

Lemma 3.46. *If v and v' are dependent then $H_v = H_{v'}$*

Proof. It is enough to show that if v depends strictly on v' then $H_v = H_{v'}$. Let v and v' be the i-th and j-th vertex respectively respect to the total order on $\mathcal{B}$ associated with certain admissible order. As $v' \in p(v)$ then $i > j$. Consider $\sigma_i : X_i \to \mathbb{P}^2$ and $\sigma_j : X_j \to \mathbb{P}^2$. Let x be the unique representative in X_{i-1} of the infinitely near point corresponding to v. If f is a generic element of $H^0(\mathbb{P}^2, \mathcal{I}(d))$ then, by the definition of strict dependence, the strict transform by σ_i of the curve defined by f does not meet $E_i \cap E_j$. This implies that the strict transform $\tilde{C}_{i-1}(f)$ of f by σ_{i-1} is not tangent to E_j at x. We denote by m the intersection multiplicity $I_x(E_j, \tilde{C}_{i-1}(f))$. The number m is determined by the graph $\mathcal{B}$.

Now we consider g in H_v. Let $\tilde{C}_i(g)$ be virtual strict transform of g by σ_i. If $\tilde{C}_i(g)$ does not contain E_i as a component then it should increase its intersection multiplicity with E_j at $E_i \cap E_j$. Hence, in any case $\tilde{C}_i(g)$ meets $E_i \cap E_j$. This implies that $I_x(E_j, \tilde{C}_{i-1}(g)) > m$, and it follows that g belongs to $H_{v'}$. Consequently $H_v \subset H_{v'}$ and hence that $H_v = H_{v'}$. □

Clearly "being dependent of" is an equivalence relation in the set of vertices of $\mathcal{B}$. It is not difficult to show that each equivalence class has an absolute maximum for the parental ordering.

DEFINITION 3.47. Define $V_{max}(\mathcal{B})$ to be the set formed by the maximal vertices of the equivalence classes of the relation "being dependent of"

PROPOSITION 3.48. *The hyperplanes* $\{H_v : v \in V_{max}(\mathcal{B})\}$ *are paiwise different and, moreover, they form a normal crossings hyperplane arrangement.*

PROOF. Denote by $v_0, ..., v_n$ the vertices of $\mathcal{B}$ ordered by the total ordering in $\mathcal{B}$. Let $\{v_{i_1}, ..., v_{i_k}\}$ be the set of vertices $V_{max}(\mathcal{B})$ ordered decreasingly by the restriction of the total ordering. We define $F_s := \cap_{j=s}^{k} H_{v_{i_j}}$. Consider the filtration $F_k \supset F_{k-1} \supset ... \supset F_1 \supset F_0 := (0)$ The proposition follows if we show that for any $1 \leq i \leq k$ there exists a polynomial $f \in F_i \setminus F_{i-1}$. An element of $F_1 \setminus F_0$ is the polynomial $Z^{\deg(\mathcal{B})}$. Consider the following claim: For any $2 \leq i \leq k$ there exists an ordered behaviour graph $\mathcal{G}_i, \leq_a$ and an ordered set $(I_i, \alpha_i) \in V_{\mathcal{G}_i, \leq_a}$ such that if $f \in \tau_{\mathcal{G}_i, \leq_a}^{-1}(I_i, \alpha_i)$ then $f \in F_i \setminus F_{i-1}$. Due to Proposition 3.31, if we prove the claim then we are done. We do it in the next lemma. □

DEFINITION 3.49. An *extended behaviour graph* is a decorated graph $\mathcal{G}$ that satisfyies conditions 1, 3, 4 and 5 of Definition 2.23 and in addition the following condition

2'. Let p be the parental function on $\mathcal{G}$. If v is any vertex then, either v satisfies condition 2 of Definition 2.23, or there exist vertices $v_1, ..., v_r$ such that $v_r = v$, $p(v_i) = \{v_1, v_{i-1}\}$ for $i \geq 3$, the vertex v_2 satisfies condition 2 and, if $i > 2$, the vertex v_i does not satisfy condition 2. If $i > 2$ we will say that v_i is a *superfluous vertex.*

REMARK 3.50. Given an extended behaviour graph $\mathcal{G}$ we can succesively delete all the superfluous vertices without modifying the decoration, getting at the end a behaviour graph $\mathcal{G}'$. Any admissible ordering $\leq_a$ in $\mathcal{G}'$ extends uniquely to an admissible ordering in $\mathcal{G}$ (which we also denote by $\leq_a$); conversely any admissible ordering in $\mathcal{G}$ restricts to a unique admissible ordering in $\mathcal{G}'$.

Define $X_{\mathcal{G}, \leq_a} := X_{\mathcal{G}', \leq_a}$. As any superfluous vertex is fixed there is a canonical isomorphism between $V_{\mathcal{G}, \leq_a}$ and $V_{\mathcal{G}', \leq_a}$, that allows us to identify these varieties; under the last identification we have a morphism $\tau_{\mathcal{G}, \leq_a} : X_{\mathcal{G}, \leq_a} \to V_{\mathcal{G}, \leq_a}$.

LEMMA 3.51. *With the notations of the proof of the last proposition, for any* $2 \leq j \leq k$ *there exists an extended behaviour graph* $\mathcal{G}_j$ *and an ordered set* $(I_j, \alpha_j) \in V_{\mathcal{G}_j, \leq_a}$ *satisfying the following properties*

(1) $\tau_{\mathcal{G}_j, \leq_a}^{-1}(I_j, \alpha_j)$ *is included in* $F_j \setminus F_{j-1}$.

(2) *The parental system of* $\mathcal{B}$ *is a saturated subset of the parental system of* $\mathcal{G}_j$*; the set of vertices* v *of* $\mathcal{G}_j \setminus \mathcal{B}$ *such that* $p(v)$ *has two elements is empty. In addition, if* $v \in \mathcal{G}_j \setminus \mathcal{B}$ *and* $v' = p(v)$ *then* v' *belongs to the dependency class of* $v_{i_{j'}}$ *for a certain* $j' \geq j$. *If* x *is either a vertex of* $\mathcal{B}$ *that is not in the dependence class of* $v_{i_{j'}}$ *for any* $j' \geq j$, *or an edge of that connects two*

vertices of $\mathcal{B}$ such that none of them is in the dependence class of $v_{i_{j'}}$ for any $j' \geq j$, then the arrows of $\mathcal{B}$ attached to x coincide with the arrows of $\mathcal{G}_j$ attached to x; moreover the corresponding weights are also preserved.

(3) *Let $\mathcal{B}_{i_j-1}$ be the graph obtained by deleting from $\mathcal{B}$ all the vertices smaller or equal than v_{i_j}. Let $\mathcal{G}_j^*$ be the graph obtained by deleting from $\mathcal{G}_j$ all the vertices that are not in $\mathcal{B}_{i_j-1}$. Then the arrows of $\Lambda_1(\mathcal{G}_j^*) \cup \Lambda_3(\mathcal{G}_j^*)$ coincide with the arrows of $\Lambda_1(\mathcal{B}_{i_j-1}) \cup \Lambda_3(\mathcal{B}_{i_j-1})$ and the weights are also preserved. Moreover the arrows of $\Lambda_2(\mathcal{G}_j^*) \cup \Lambda_4(\mathcal{G}_j^*)$ are the same as the arrows of $\Lambda_2(\mathcal{B}_{i_j-1}) \cup \Lambda_4(\mathcal{B}_{i_j-1})$ and the primary components of their weights are also preserved.*

PROOF. We prove the lemma by descending induction on j. The initial and the inductive steps follow the same pattern, and hence we only take care of the latter. Assume that the lemma is true for any $j' > j$, we show it for j. We split our reasoning into different cases depending on the nature of v_{i_j}.

Case 1. The vertex v_{i_j} is a fixed (lateral or central) vertex. Let $p(v_{i_j}) = \{v'_1, v'_2\}$, where v'_1 is a free central vertex, (note that whenever a vertex is fixed, one of its parents is free central). To simplify the notation we denote v_{i_j} by w_1. Let $w_2, ..., w_r$ be the maximal sequence in the dependence class of v_{i_j} such that $p(w_k) = \{v'_1, w_{k-1}\}$ if $k > 1$ (i.e. the maximal sequence of fixed vertices of the dependence class). The following two claims hold:

(1) There is an arrow a in $\mathcal{G}_{j+1}$ such that $\gamma(a) = \{v'_2, w_1\}$.
(2) There is a vertex w_{r+1} in $\mathcal{G}_{j+1}$ such that $p(w_{r+1}) = \{v'_1, w_r\}$.

The arrow a exists as an arrow of $\mathcal{B}$ since otherwise w_1 would not be maximal in its dependence class. As neither v'_2 nor w_1 belong to the dependence class of any $v_{i_{j'}}$ with $j' \geq j+1$ we deduce that a exists as an arrow of $\mathcal{G}_{j+1}$. Suppose that the vertex of claim 2 does not exist. Then there is no arrow b in $\mathcal{G}_{j+1}^*$ such that $\gamma(b) = \{v'_1, w_r\}$. As the arrows of $\mathcal{G}_{j+1}^*$ and $\mathcal{B}_{i_{j+1}-1}$ are the same we deduce that there is no arrow b in $\mathcal{B}_{i_{j+1}-1}$ such that $\gamma(b) = \{v'_1, w_r\}$. This implies that w_r and v'_1 are strictly dependent, which is in contradiction with the fact that w_1 is the maximal element of its dependence class. Let $\{w'_1, ..., w'_s = v'_2\}$ be the unique sequence such that $p(w_1) = \{v'_1, w'_s\}$, $p(w'_k) = \{v'_1, w'_{k-1}\}$ for $1 < k \leq s$ and $p(v'_1) = \{w'_1\}$ (the number s is the distance of w_1 to the beginning of the floor where it is located).

We have to consider two cases, depending on whether w_r and w_{r+1} are in the same dependence class or not viewed as vertices of $\mathcal{G}_{j+1}$.

Subcase 1.1. Assume that w_r and w_{r+1} are not in the same dependence class as vertices of $\mathcal{G}_{j+1}$. In this case there should exist an arrow a' in $\mathcal{G}_{j+1}$ such that $\gamma(a') = \{w_r, w_{r+1}\}$. Let (e, f) and (e', f') be the respective weights of a and a'. We construct the graph $\mathcal{G}_j$ by modifying $\mathcal{G}_{j+1}$ in the following way:

- If $e \neq 1$ and $f' \neq 1$ then we change the weights of a, a' to $(e-1, f+1)$, $(e', f'-1)$ respectively. We add vertices $\nu'_1, ..., \nu'_{s+r+1}$ with parental function $p(\nu'_{k+1}) = \nu'_k$ for $k > 1$ and $p(\nu'_1) = w_r$, and attach an arrow a''' to ν'_{s+1} with weight 1.
- If $e = 1$ and $f' \neq 1$ we consider the graph constructed in the last item, delete from it the arrow a, add vertices $\nu_1, ..., \nu_{s+1}$ with parental function $p(\nu_{k+1}) = \nu_k$ for $k > 1$ and $p(\nu_1) = w_1$, and attach an arrow a'' to ν_{s+1} with weight $f + 1$.

- If $e \neq 1$ and $f' = 1$ we consider the graph constructed in the first item, delete from it the arrow a', and change the weight of a''' to $e' + 1$.
- If $e = 1$ and $f' = 1$ we consider the graph constructed in the second item, delete from it the arrow a' and change the weights of a'', a''' to $f + 1$, $e' + 1$ respectively.

To define I_j we impose that if v is a vertex of $\mathcal{G}_j$ that was already present in $\mathcal{G}_{j+1}$, then the infinitely near point of I_j associated with v is the point of I_{j+1} that corresponds to v viewed as a vertex of $\mathcal{G}_{j+1}$. To define the infinitely near point associated with ν_1 we choose any point of $\dot{E}_{w_1}$ that is not in I_{j+1}. For $k > 1$ we define the infinitely near point associated with ν_k choosing any point of $\dot{E}_{\nu_{k-1}}$. The infinitely near points associated with the ν_k''s are defined in an analogous way. The vertices $\nu_1, ..., \nu_{s+1}$ are totally ordered by the parental order, and the same holds for $\nu_1', ..., \nu_{s+r+1}'$. As ν_1 and ν_1' are attached to different vertices of $\mathcal{G}_{j+1}$ there exists a unique total order $\leq_t$ in $\mathcal{G}_j$ that comes from an admissible ordering $\leq_a$ and extends the total order of $\mathcal{G}_{j+1}$ induced by α_{j+1}. To define an element of $V_{\mathcal{G}_j, \leq_a}$ is equivalent to give an element of $V_{\mathcal{G}_j}$ together with a certain total order on it that extends the canonical order. We define (I_j, α_j) to be I_j together with the total ordering $\leq_t$.

If the dependence class of w_{r+1} in $\mathcal{G}_{j+1}$ consists of $\{w_{r+1}, v_1'\}$, the weight of a' is $(e', 1)$ and there is no vertex v such that $p(v) = \{w_{r+1}\}$, then $\mathcal{G}_j$ is an extended behaviour graph, and w_{r+1} is a superfluous vertex. This is the only case where extended behaviour graphs arise.

By induction we know that $\mathcal{G}_{j+1}$ satisfies the second property of the lemma. Then, by construction, it is clear that $\mathcal{G}_j$ also does.

In $\mathcal{G}_j^*$ the vertices v_1' and w_s' are connected by an edge since $p(v_{i_j}) = p(w_1) = \{v_1', w_s'\}$. Any arrow a of $\mathcal{G}_j^*$ not attached to the edge connecting v_1' with w_s is also an arrow of $\mathcal{G}_{j+1}^*$. By the induction hypothesis, such an arrow a satisfies the third condition of the lemma. In $\mathcal{B}_{i_j-1}$ there is an arrow b attached to the edge connecting v_1' with w_s, that comes from the deletion of v_{i_j}. Let its weight be $(t_1, t_2, t_3, t_4, t_5)$. In $\mathcal{G}_j^*$ there is an arrow b' attached to the same edge, with weight $(t_1, t_2, t_3 + 1, t_4 + 1, t_5)$ or $(t_1, t_2, t_3 + 1, t_4, t_5 + 1)$, depending on whether w_s is the first vertex of its floor or not. In any case, the primary components of the weights of b and b' are the same. This finishes the proof of the third condition of the lemma.

It remains to show the first condition of the lemma. Consider $f \in \tau_{\mathcal{G}_j, \leq_a}^{-1}(I_j, \alpha_i)$. We claim that $f \in H^0(\mathbb{P}^2, \mathcal{I}_{I_{j+1}}(d))$. To prove the claim we have to show that $f \in H^0(\mathbb{P}^2, \mathcal{I}_{I_{j+1},k}(d))$ for each vertex v_k of $\mathcal{G}_{j+1}$, (where $\mathcal{I}_{I_{j+1},k}$ is defined in construction 3.23). If $v_k >_t v_{i_j}$, the arrows corresponding to the deletion of v_k in $\mathcal{G}_j$ and $\mathcal{G}_{j+1}$ have both the same weight than the arrow corresponding to the deletion of v_k in $\mathcal{B}$. This is because these weights we only depend on the primary components of the weights of the arrows of $\mathcal{G}_j^*$ and $\mathcal{G}_{j+1}^*$, which are equal to the primary components of the weights of the respective arrows of $\mathcal{B}_{i_j-1}$. Thus $\mathcal{I}_{I_j,k} = \mathcal{I}_{I_{j+1},k}$ for any $k < i_j$. Since f belongs to $\tau_{\mathcal{G}_j, \leq_a}^{-1}(I_j, \alpha_j)$ it is in $H^0(\mathbb{P}^2, \mathcal{I}_{I_j,k}(d))$ for any $k < i_j$ and hence f is in $H^0(\mathbb{P}^2, \mathcal{I}_{I_{j+1},k}(d))$.

Consider $v_k = v_{i_j}$, assume for simplicity that w_s is the first vertex of its floor. If $(t_1, t_2, t_3, t_4, t_5)$ is the weight of the arrow corresponding to the deletion of v_k in $\mathcal{G}_{j+1}$, then the weight of the arrow corresponding to the deletion of v_k in $\mathcal{G}_j$ is $(t_1, t_2, t_3 + 1, t_4 + 1, t_5)$. Hence the conditions imposed by $\mathcal{I}_{I_j,k}$ imply the conditions imposed by $\mathcal{I}_{I_{j+1},k}$. Consider an affine chart $(V,(x,y))$ centered in x_{i_j} such that $E_{v_1'}$ and $E_{v_2'}$ are defined by $y=0$ and $x=0$ respectively. Let g be the expression of the virtual strict transform of f (viewed as an element of $H^0(\mathbb{P}^2,\mathcal{I}_{I_{j+1}}(d))$) by σ_{i_j-1} with respect to the coordinates (x,y). An easy argument using the Newton polygon of g shows that if $v_k=w_l$ for $1\leq l\leq r+1$ then the conditions imposed by $\mathcal{I}_{I_j,k}$ imply the conditions imposed by $\mathcal{I}_{I_{j+1},k}$. If v_k is any other vertex of $\mathcal{G}_{j+1}$ then the weight of the arrow corresponding to the deletion of v_k in $\mathcal{G}_{j+1}$ coincides with the weight of the arrow corresponding to the deletion of v_k in $\mathcal{G}_j$, on the other hand it is a combinatorial exercise to check that the virtual multiplicity of any vertex of $\mathcal{G}_{j+1}$ is smaller or equal than the virtual multiplicity of the same vertex viewed in $\mathcal{G}_j$. This implies that $\mathcal{I}_{I_j,k}\subset\mathcal{I}_{I_{j+1},k}$. Therefore

$$\mathcal{I}_{I_j}=\cap_{v_k\in\mathcal{G}_j}\mathcal{I}_{I_j,k}\subset\cap_{v_k\in\mathcal{G}_{j+1}}\mathcal{I}_{I_{j+1},k}=\mathcal{I}_{I_{j+1}}$$

and hence that $H^0(\mathbb{P}^2,\mathcal{I}_{I_j}(d))\subset H^0(\mathbb{P}^2,\mathcal{I}_{I_{j+1}}(d))$. The claim is proved.

Next we show that any f in $\tau^{-1}_{\mathcal{G}_j,\leq a}(I_j)$ belongs to F_j: the claim above implies that $f\in F_{j+1}$. As $F_j=F_{j+1}\cap H_{v_{i_j}}$ we only have to show that $f\in H_{v_{i_j}}$, and this follows from the fact that the virtual multiplicity of v_{i_j} in $\mathcal{G}_j$ is strictly bigger than the virtual multiplicity of v_{i_j} in $\mathcal{G}_{j+1}$. To show that f is not in F_{j-1} we have to show that f is not in $H_{v_{j-1}}$. Let $v\in\mathcal{G}_j$ be such that $v\geq_p v_{i_j-1}$. As $v_{i_{j-1}}\in\mathcal{G}_j^*$, using the third property, the arrow corresponding to the deletion of v as a vertex of $\mathcal{G}_j$ has the same weight than the arrow corresponding to its deletion as a vertex of $\mathcal{B}$. As $f\in\tau^{-1}_{\mathcal{G}_j,\leq a}(I_j,\alpha_j)$, this implies that f is not in $H_{v_{j-1}}$ and concludes the proof of this subcase.

In the subsequent cases we will only define $\mathcal{G}_j$ and (I_j,α_j); the fact that such definitions work can be proved in the same spirit than in last case.

Subcase 1.2. Assume that w_r and w_{r+1} are in the same dependence class as vertices of $\mathcal{G}_{j+1}$. As w_{r+1} is a fixed vertex the maximal vertex v_{i_h} in its dependence class as vertices of $\mathcal{B}$ should be either w_{r+1} or v_1'. Suppose that $v_{i_h}=v_1'$. As w_r and w_{r+1} are in different dependency classes as vertices of $\mathcal{B}$ (otherwise the sequence $w_1,\dots,w_r$ would not be maximal) we know that there is an arrow b in $\mathcal{B}$ such that $\gamma(b)=\{w_r,w_{r+1}\}$. As $v_1'>_p w_1$ we have that, for any $j'\geq j+1$, neither w_r nor w_{r+1} belong to the dependency class of $v_{i_{j'}}$ as vertices of $\mathcal{B}$. By induction hypothesis, using property 2, we deduce that b should be also an arrow of $\mathcal{G}_{j+1}$. This contradicts the fact that w_r and w_{r+1} are in the same dependence class as vertices of $\mathcal{G}_{j+1}$; hence $v_{i_h}=w_{r+1}$.

As w_r and w_{r+1} are in the same dependence class as vertices of $\mathcal{G}_{j+1}$ the arrow b is not present in $\mathcal{G}_{j+1}$. It only can disappear when we modify $\mathcal{G}_{h+1}$ to construct $\mathcal{G}_h$. In this case its weight should be of the form $(1,g)$. This implies that when we pass from $\mathcal{G}_{h+1}$ to $\mathcal{G}_h$ we delete b, add vertices $\nu_1'',\dots,\nu_{s+r+2}''$, with parental function $p(\nu_1')=w_{r+1}$ and $p(\nu_{k+1}'')=\nu_k'$ for $k>1$, and attach an arrow b' to ν_{s+r+2}' with weight $g+1$. Recall that in $\mathcal{G}_{j+1}$ there is an arrow a such that $\gamma(a):=\{w_s',w_1\}$ with weight (e,f). We construct $\mathcal{G}_j$ by modifying $\mathcal{G}_{j+1}$ in the following way:

- If $e\neq 1$ then we change the weights of a, b' to $(e-1,f+1)$, g respectively. We add vertices $\nu_1',\dots,\nu_{s+r+1}'$ with parental function $p(\nu_{k+1}')=\nu_k'$ for $k>1$ and $p(\nu_1')=w_r$, and attach an arrow a''' to ν_{s+r+1}' with weight 1.
- If $e=1$ we do as in the last item but instead to changing the weight of a we delete it, add vertices $\nu_1,\dots,\nu_{s+1}$ with parental function $p(\nu_{k+1})=\nu_k$

for $k > 1$ and $p(\nu_1) = w_1$, and attach an arrow a'' to ν_{s+1} with weight $f+1$.

To define (I_j, α_j) we only have to specify the infinitely near points corresponding to the newly added vertices. This is analogous to the last subcase.

It is clear that we can deal with the case in which v_{i_h} is of subcase 1.2 in an exactly analogous manner.

Case 2. The vertex v_{i_j} is a free lateral vertex. Consider the parental system formed for all the vertices of $\mathcal{B}$ bigger or equal than v_{i_j} by the parental ordering, Let $\mathcal{A}$ be its graph. We denote by r the length of the last floor of $\mathcal{A}$ and assume that v_{i_j} occupies the $s-th$ position in the free lateral vertices of this last floor (where $s \leq r-1$). Let $p(v_{i_j}) = v_k$. There is an arrow a such that $\gamma(a) = \{v_k, v_{i_j}\}$, because otherwise v_{i_j} would not be maximal in its dependence class. Let (e, f) be its weight. We define $\mathcal{G}_j$ by modifying $\mathcal{G}_{j+1}$ in the following way:

- If $e \neq 1$ change the weight of a to $(e-1, f)$ and add a sequence of vertices $\nu_1, ..., \nu_{r-s}$ such that $p(\nu_1) = v_{i_j}$ and $p(\nu_k) = \nu_{k-1}$ if $k > 1$. Attach to ν_{r-s} an arrow a' with weight 1
- If $e = 1$ delete a and add a sequence of vertices $\nu_1, ..., \nu_{r-s}$ and an arrow a' as above but with weight $f+1$.

Define I_j by adding to I_{j+1} the set x_1,...,x_{s-r}, where x_1 is an infinitely near point of $\dot{E}_{i_j}$ that is not in I_{j+1} and for $k > 1$ we choose x_k to be any point of $\dot{E}_{\nu_{k-1}}$. The isomorphism α_j is defined in a way analogous to the preceding cases.

Case 3. The vertex v_{i_j} is free central and there is an arrow b attached to v_{i_j}. Let c be the weight of b. Let w be the last vertex such that $v_{i_j} \in p(w)$, consider the parental system formed for all the vertices of $\mathcal{B}$ bigger or equal than w by the parental ordering, let $\mathcal{A}$ be its graph. The vertex v_{i_j} is the free central vertex of the last floor of $\mathcal{A}$. We denote by r the length of the last floor. There is at least an arrow attached to an edge that meets a fixed lateral vertex, because otherwise the first vertex of the floor (i.e. $p(v_{i_j})$) would be in the same dependence class of v_{i_j} and this would contradict the fact that v_{i_j} is maximal in the parental ordering within its class. Let v_1 be the first vertex and $v_2, ..., v_r = w$ the fixed vertices of the last floor of $\mathcal{A}$. Let v_s be the last of them such that there is an arrow a in $\mathcal{B}$ such that $\gamma(a) = \{v_{s-1}, v_s\}$. The dependence class of v_{i_j} includes $v_{i_j}, v_s, ..., v_r$. To define $\mathcal{G}_j$ we need to consider two different subcases depending on wether v_{s-1} and v_s are in the same dependence class as vertices of $\mathcal{G}_{j+1}$ or not.

Subcase 3.1. Assume that v_{s-1} and v_s are not in the same dependence class as vertices of $\mathcal{G}_{j+1}$. In this case the arrow a has not been deleted in the construction of $\mathcal{G}_{j+1}$, we let (e, f) be its weight. We construct $\mathcal{G}_j$ by modifying $\mathcal{G}_{j+1}$ in the following way:

- If $e \neq 1$ change the weight of a to $(e-1, f+1)$. Change the weight of b to be $c-1$, and if $c = 1$ then delete b.
- If $e = 1$ delete a and add a sequence of vertices $\nu_1, ..., \nu_s$ such that $p(\nu_1) = v_s$ and $p(\nu_k) = \nu_{k-1}$ if $k > 1$. Attach to ν_{r-s} an arrow a' with weight $f+1$. Change the weight of b to be $c-1$, and if $c = 1$ then delete b.

If $e \neq 1$ we let $I_j := I_{j+1}$ and if $e = 1$ we define I_j by adding to I_{j+1} the set x_1,...,x_{s-1} where the x_i's are defined analogously than in the case above. The definition of the isomorphism α_j is also analogous.

Subcase 3.2 Assume that v_{s-1} and v_s are in the same dependence class as vertices of $\mathcal{G}_{j+1}$. The arrow a does not appear in $\mathcal{G}_{j+1}$. Consider the maximal vertex v_{i_h} of the dependence class of w_{s-1} in $\mathcal{B}$. The only moment in which b could have been deleted is when passing from $\mathcal{G}_{h+1}$ to $\mathcal{G}_h$, and in this case v_{i_h} should be of case 1 and the weight of a of the form $(e, 1)$. In the course of the construction of $\mathcal{G}_h$ we add a sequence of vertices $\nu'_1, ...\nu'_{s-1}$ (with parental function $p(\nu'_1) = v_{s-1}$ and $p(\nu'_k) = \nu'_{k-1}$ for $k > 1$) and an arrow a'' attached to ν'_{s-1} and with weight $e+1$. These vertices and arrow appear also in $\mathcal{G}_{j+1}$. We construct the graph $\mathcal{G}_j$ by modifying $\mathcal{G}_{j+1}$ in the following way:

- Change the weight of b to be $c-1$, and if $c = 1$ then delete b. Decrease the weight of a'' in one unit and add a sequence of vertices $\nu_1, ..., \nu_s$ such that $p(\nu_1) = v_s$ and $p(\nu_k) = \nu_{k-1}$ if $k > 1$. Attach to ν_{r-s} an arrow a' with weight 1.

The set I_j and the automorphism α_j are constructed analogously to the cases above.

Case 4. The vertex v_{i_j} is free central and there is no arrow b attached to v_{i_j}. We let $\mathcal{A}$ and $v_1, ..., v_s, ..., v_r$ be like in last case. Again we have to treat two subcases depending on whether v_s and v_{s-1} are in the same dependence class as vertices of $\mathcal{G}_{j+1}$.

Subcase 4.1 The vertices v_s and v_{s-1} are not in the same dependence class as vertices of $\mathcal{G}_{j+1}$. The same argument than before show that there should exist an arrow in $\mathcal{G}_{j+1}$ attached to the edge connecting v_s with v_{s-1}. Let the weight of a be (e, f). We define $\mathcal{G}_j$ by modifying $\mathcal{G}_{j+1}$ in the following way:

- If $e \neq 1$ change the weight of a to $(e-1, f)$ and attach an arrow to v_{i_j} with weight $s-1$.
- If $e = 1$ delete a and add a sequence of vertices $\nu_1, ..., \nu_s$ such that $p(\nu_1) = v_s$ and $p(\nu_k) = \nu_{k-1}$ if $k > 1$. Attach to ν_s an arrow a' with weight f. Attach an arrow to v_{i_j} with weight $s-1$.

The ordered set (I_j, α_j) is defined like in last case.

Subcase 4.2 The vertices v_s and v_{s-1} are in the same dependence class as vertices of $\mathcal{G}_{j+1}$. Consider $\nu'_1, ...\nu'_{s-1} \in \mathcal{G}_{j+1}$ and $a'' \in \mathcal{G}_{j+1}$ like in subcase 3.2. We construct the graph $\mathcal{G}_j$ by modifying $\mathcal{G}_{j+1}$ in the following way:

- We attach an arrow to v_{i_j} with weight $s-1$ and decrease in a unit the weight of a''

The ordered set (I_j, α_j) is constructed analogously to the cases above. The proof of this case ends the proof of the lemma. □

NOTATION 3.52. Let $\mathcal{B}$ be any behaviour graph. Denote by $\mathcal{S}_1$ and $\mathcal{S}_2$ the sets of vertices v such that $p(v)$ has 1 and 2 vertices respectively. Let $\leq_a$ be an admissible ordering. Given any vertex v consider the graph $\mathcal{B}_v$ obtained by deleting all the vertices that are smaller or equal than v in the total ordering associated to $\leq_a$. In $\mathcal{B}_v$ there is an arrow a_v associated with v. If $v \in \mathcal{S}_1$ (resp. $v \in \mathcal{S}_2$) let $(t_1(v), t_2(v), t_3(v))$ (resp. $(t_1(v), t_2(v), t_3(v), t_4(v), t_5(v))$) be the weight of a_v.

DEFINITION 3.53. Let $\mathcal{B}, \leq_a$ be an ordered behaviour graph. Define

$$\begin{array}{lll} \epsilon(\mathcal{B}) & := & \text{number of dependency classes} \\ \delta_1(\mathcal{B}) & := & \text{number of free vertices} \\ \delta_2(\mathcal{B}) & := & (\sum_{v\in\mathcal{S}_1}(t_1(v)+\frac{(t_2(v)-1)t_2(v)}{2}+\max(0,t_3(v)-1))+ \\ & & (\sum_{v\in\mathcal{S}_2}(\frac{(t_3(v)-2)(t_3(v)-1)}{2}+\max(0,t_4(v)-1)+\max(0,t_5(v)-1)) \end{array}$$

Consider the sequence $v_{h_1},...,v_{h_{\delta_1(\mathcal{B})}}$ of all the free vertices ordered by the associated total ordering. Define, for any $2 \leq i \leq \delta_1(\mathcal{B})$

$$\xi_i := -\chi(\dot{E}_{p(v_{h_i})}) + 1 - \sharp\{v : v \leq_t v_{h_i},\ p(v) = p(v_{h_i})\}$$

The next theorem explains the geometry of $\tau_{\mathcal{B},\leq_a}$:

THEOREM 3.54. *Let $\mathcal{B}, \leq_a$ be any ordered behaviour graph. The morphism $\tau_{\mathcal{B},\leq_a} : X_{\mathcal{B},\leq_a} \to V_{\mathcal{B},\leq_a}$ is surjective, smooth of relative dimension*

$$(d+1)(d+2)/2 - \delta_2(\mathcal{B})$$

and a locally trivial fibration whose fibres are isomorphic to

$$\mathbb{C}^{(d+1)(d+2)/2-\delta_2(\mathcal{B})-\epsilon(\mathcal{B})} \times (\mathbb{C}^*)^{\epsilon(\mathcal{B})}$$

The variety $X_{\mathcal{B},\leq_a}$ is smooth and rational.

The variety $V_{\mathcal{B},\leq_a}$ is smooth of dimension $\delta_1(\mathcal{B})$. There is a sequence

$$V_{\mathcal{B},\leq_a} := V_{\delta_1(\mathcal{B})} \xrightarrow{\pi_{\delta_1(\mathcal{B})}} V_{\delta_1(\mathcal{B})-1} \to ... \to V_1 \tag{3.12}$$

of fibrations such that $V_1 \cong \mathbb{P}^1$ and for any $2 \leq i \leq \delta_1(\mathcal{B})$ each fibre of π_i is $\mathbb{C} \setminus \{\xi_i$ points$\}$.

PROOF. As $V_{\mathcal{B},\leq_a}$ is a smooth variety over $\mathbb{C}$, and $\tau_{\mathcal{B},\leq_a}$ is equidimensional with smooth fibres, then it is smooth. The fibres are as claimed because, for any $(I,\alpha) \in V_{\mathcal{B},\leq_a}$, the codimension of $H^0(\mathbb{P}^2, \mathcal{I}(d))$ in $H^0(\mathbb{P}^2, \mathcal{O}_{\mathbb{P}^2}(d))$ is $\delta_2(\mathcal{B})$ (as one deduces from the formulae 3.8 and 3.9), and the fibre of $\tau_{\mathcal{B},\leq_a}$ over (I,α) consists of deleting from $H^0(\mathbb{P}^2, \mathcal{I}(d))$ exactly $\epsilon(\mathcal{B})$ linearly independent hyperplanes (see Proposition 3.48).

The sequence 3.12 is the sequence of Corollary 3.15. □

All the morphisms of the sequence

$$X_{\mathcal{B},\leq_a} \xrightarrow{\tau_{\mathcal{B},\leq_a}} V_{\delta_1(\mathcal{B})} \xrightarrow{\pi_{\delta_1(\mathcal{B})}} V_{\delta_1(\mathcal{B})-1} \to ... \to V_1$$

are C^∞ locally trivial fibrations with fibres are spaces whose i-th homotopy group vanishes for any $i > 1$. By the long homotopy sequence of a fibration, and because the higher homotopy groups of a space are isomorphic to the higher homotopy groups of any covering of it we get

COROLLARY 3.55. *Let $\mathcal{B}$ be any behaviour graph. If $i \geq 3$ then the i-th homotopy group of $X_{\mathcal{B},\leq_a}$ and $V_{\mathcal{B},\leq_a}$ are*

$$\pi_i(X_{\mathcal{B},\leq_a}) \cong \pi_i(V_{\mathcal{B},\leq_a}) \cong \pi_i(X_{\mathcal{B}}) \cong \pi_i(V_{\mathcal{B}}) \cong \pi_i(S^2)$$

where S^2 is the sphere of dimension 2.

CHAPTER 4

The action of $Aut(\mathbb{C}^2)$ on $\mathbb{C}[x, y]$

In this chapter we study the action $Aut(\mathbb{C}^2)$ in $\mathbb{C}[x, y]$ by means of the behaviour graph. In the first section we study how the behaviour graph of a polynomial changes under composition with an automorphism; this leads naturally to the definition of an equivalence relation in the set of behavior graphs. Using of the results of the first section, in the second section we study how the degree of a polynomial changes by composition with an automorphism. Finally, in the third section, we show the unexpected fact that, for any $f \in \mathbb{C}[x, y]$ and $t \in \mathbb{C}$, both f and $f + t$ have the same behaviour graph. This result is known not to be true for embedded resolutions graphs of the singularities at infinity; the definition of behaviour graph plays a fundamental role in its proof.

4.1. The action of $Aut(\mathbb{C}^2)$ at the combinatorial level

Consider the rooted weighted graph $\mathcal{A}_n$. Let $\{v_0, ..., v_{2n-1}\}$ be its set of vertices ordered by the parental order. By the symmetry of the graph, the rooted graph obtained from $\mathcal{A}_n$ letting the root become v_{2n-1} is isomorphic to $\mathcal{A}_n$. Then the parental ordering would order the vertices as $\{v_{2n-1}, v_1, v_{2n-2}, ..., v_n, v_{n-1}, ..., v_2, v_0\}$; the last vertex becomes the root; the root becomes the last vertex; the free and fixed central vertices keep being the free and fixed central vertices; the free lateral vertices except the last one, become the fixed lateral vertices, and they invert their ordering; finally the fixed lateral vertices become free lateral vertices also inverting their ordering. If $\mathcal{A}_n$ is the graph of a primitive automorphism, then the graph obtained by letting the last vertex become the root is the graph of the inverse of the primitive automorphism.

Consider the rooted weighted graph $\mathcal{A}_{n_1,...,n_r}$. If we let the last vertex become the root then we obtain the rooted weighted graph $\mathcal{A}_{n_r,...,n_1}$. By this operation the i-th floor of $\mathcal{A}_{n_1,...,n_r}$ becomes the $r - i$-th floor of $\mathcal{A}_{n_r,...,n_1}$, and at each floor the operation changes the root into the last vertex. It is clear that if $\mathcal{A}_{n_1,...,n_r}$ is the graph of an automorphism, the graph obtained by letting the last vertex become the root is the graph of the inverse of that automorphism.

Suppose that we are given:

(1) A decorated graph $\mathcal{G}$ whose parental system is called (I, p_I).
(2) A parental system (S, p) whose graph is $\mathcal{A}_{n_1,...,n_r}$ for certain $r, n_1, ..., n_r$.
(3) A parental system (S_0, p_0) identified at the same time with a subset of I and with a subset of S, with the following two properties: first, S_0 contains the root of both (I, p_I) and (S, p), and p_0 is the restriction of p_I (resp. p) to S_0, and second, if $x \in I$ and $x' \in S$ are such that $p_I(x) = \{y, z\}$ and $p(x') = \{y, z\}$ with $\{y, z\} \subset S_0$ then there exist $x'' \in S_0$ such that x'' is identified with x in I and with x' in S.

DEFINITION 4.1. A triple $\mathcal{G}, (S,p), (S_0, p_0)$ as above is called a composable triple.

Given $x, y \in I \backslash S_0$, we say that x is connected to y in $I \backslash S_0$ if there exist $z \in I \backslash S_0$ such that $x \leq_p z$ and $y \leq_p z$. "Being connected to" defines an equivalence relation in $I \setminus S_0$; reflexiveness and symmetry are obvious, transitivity is a consequence of the following property of $\leq_p$: if x, y, z are elements of I such that $x \geq_p z$ and $y \geq_p z$ then either $x \geq_p y$ or $y \geq_p x$. Another consequence of this property is that each equivalence class by the equivalence relation has a global maximum by $\leq_p$. We will denote the equivalence classes by $B_1, ... B_s$, and their respective maxima by $m_1, ..., m_s$. We say that B_i is *simply* (resp. *doubly*) *rooted* if and only if the cardinality of $p_I(m_i)$ is 1 (resp. 2). In the following definition we need to generalize bit our notion of rooted graph to have also "doubly rooted" graphs. As it will become obvious how to do it we do not give a formal definition of this.

DEFINITION 4.2. Given $\mathcal{G}$, (S,p), (S_0, p_0) a composable triple, let $B_1, ... B_s$ be the equivalence classes of $I \setminus S_0$ by the relation 'being connected to', and $m_1, ..., m_s$ their respective maxima. For each B_i we define its *associated branch* of $\mathcal{G}$ as the graph whose set of vertices is $B_i \cup p_I(m_i)$, with the same weights as if they were viewed as vertices of $\mathcal{G}$, whose edges are the edges of $\mathcal{G}$ connecting vertices of $B_i \cup p_I(m_i)$, and whose decoration is the set of arrows of $\mathcal{G}$ that are attached to vertices or edges of the branch. We denote it by $\mathcal{B}_i$. We define the *root* of $\mathcal{B}_i$ to be the set $p(m_i)$. When B_i is simply rooted then we say that the branch $\mathcal{B}_i$ is *simply rooted*, when B_i is doubly rooted then we say that $\mathcal{B}_i$ is *doubly rooted*.

We construct a rooted decorated graph as follows: define S_0' to be the union of the $p(m_i)$'s, define S_0'' to be the subset of vertices of S_0 which, viewed as vertices of $\mathcal{G}$, are decorated or meet a decorated edge. Consider in S the parental function p' that arises letting the last vertex of $\mathcal{A}_{n_1,...,n_r}$ become the root, and let $\leq_{p'}$ be its associated parental order. As S is totally ordered by $\leq_{p'}$ the set $S_0' \cup S_0''$, regarded as a subset of (S, p') has a global minimum m. Define

$$S_1 := \{q \in S \quad \text{such that} \quad q \geq_{p'} m\} \tag{4.1}$$

The pair (S_1, p') is a parental system. Let $\mathcal{G}_1$ be the associated graph. We decorate $\mathcal{G}_1$ as follows: if v is a vertex of S_1 that is not in the root of a branch then we let it carry the same decoration that it had as a vertex of $\mathcal{G}$; if e is an edge that is not connecting two vertices that are the roots of a doubly rooted branch then we let it carry the same decoration that it had as an edge of $\mathcal{G}$; we do not add decoration to the remaining vertices or edges. Now we insert the branches $\{\mathcal{B}_1, ..., \mathcal{B}_s\}$ as follows:

- If $\mathcal{B}_i$ is simply rooted and w_i is the root we notice that w_i may be viewed also as a vertex of $\mathcal{G}_1$. Let ω_i be the weight of w_i as a vertex of $\mathcal{G}_1$. We add $\mathcal{B}_i$ identifying w_i as a vertex of $\mathcal{G}_1$ with w_i as a vertex of $\mathcal{B}_i$. Let d be the number of vertices of $\mathcal{B}_i$ between w_i and m_i. We weight w_i with the number $\omega_i - d - 1$.
- If $\mathcal{B}_i$ is doubly rooted and w_j, w_l are the vertices that form their root we notice that they can be also viewed as vertices of $\mathcal{G}_1$, and that there they are connected by an edge e. Let ω_j,ω_k be their respective weights as vertices of $\mathcal{G}_1$. We delete e and identify w_j (resp w_l) as a vertex of $\mathcal{G}_1$ with w_j (resp. w_l) as a vertex of $\mathcal{B}_i$. Let d_j (resp. d_l) be the number of vertices of $\mathcal{B}_i$ between w_j (resp. w_l) and m_i. We weight w_i (resp. w_l) with the number $\omega_j - d_j - 1$ (resp. $\omega_l - d_l - 1$).

After glueing all the branches we obtain a decorated graph that we denote by $\mathcal{G}_c$.

Suppose that $\mathcal{G}$ came with an admissible ordering $\leq_a$. We are going to construct an admissible order $\leq_a'$ in $\mathcal{G}_c$ out of $\leq_a$; for this we need a way of comparing the roots of the different branches: let $\mathcal{B}_i$, $\mathcal{B}_j$ branches of $\mathcal{G}_c$, and m_i, m_j their maximal vertices, if $p(m_i) = \{v_i\}$, $p(m_j) = \{v_j\}$, then $p(m_i) < p(m_j)$ if $v_i <_{p'} v_j$; if $p(m_i) = \{v_i, v_i'\}$, $p(m_j) = \{v_j, v_j'\}$ with $v_i >_p v_i'$, $v_j >_p v_j'$, then $p(m_i) < p(m_j)$ if $v_i <_{p'} v_j$; if $p(m_i) = \{v_i\}$, $p(m_j) = \{v_j, v_j'\}$, with $v_j >_p v_j'$, then $p(m_i) < p(m_j)$ if $v_i <_{p'} v_j'$ and $p(m_i) > p(m_j)$ if $v_i \geq_{p'} v_j'$. Assume that v, v' are two vertices of $\mathcal{G}_c$ that are in the same combinatorial class:

- If $v \in \mathcal{B}_i$ and $v' \in \mathcal{B}_j$ with $p(m_i) = p(m_j)$, then they are in the same combinatorial class viewed as vertices of $\mathcal{G}$, and hence can be compared by the order $\leq_a$. We say that $v \leq_a' v'$ if $v \leq_a v'$.
- If $v \in \mathcal{B}_i$ and $v' \in \mathcal{B}_j$ with $p(m_i) < p(m_j)$ then $v \leq_a' v'$.
- If v belongs to S and v' does not, we impose $v \leq_a' v'$.
- If both v and v' belong to S, then, as they are $\sim_c$-equivalent and S is ordered linearly by the parental order, they have to be equal.

Assume that two a, a' arrows are in the same combinatorial class of $\mathcal{G}_c$:

- If $\gamma(a) = \gamma(a')$ they are in same combinatorial class as arrows of $\mathcal{G}$, and can be compared by $\leq_a$. We say that $a \leq_a' a'$ if $a \leq_a a'$.
- If $\gamma(a) \neq \gamma(a')$ then the ordering among the arrows a and a' is given by the ordering $\leq_a'$ at vertices and the properties of admissible orderings.

DEFINITION 4.3. Given $\mathcal{G}$, (S,p), (S_0,p_0), a composable triple, we define $\mathcal{G}_c$ to be the (S_0,p_0)-*composition of* $\mathcal{G}$ *and* (S,p). If $(\mathcal{G}, \leq_a)$ is an ordered graph, then we define $(\mathcal{G}, \leq_a')$ (as constructed in the last discussion) to be the (S_0,p_0)-*composition of* $(\mathcal{G}, \leq_a)$ *and* (S,p).

EXAMPLE 4.4. Let $\mathcal{G}$ be the following decorated graph

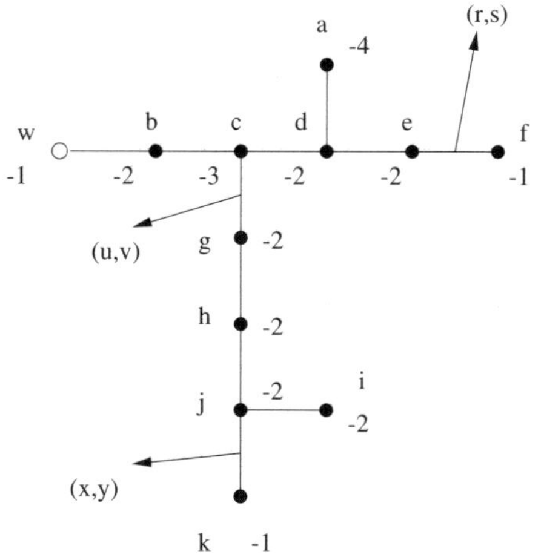

Let (S,p) be the parental system whose graph is

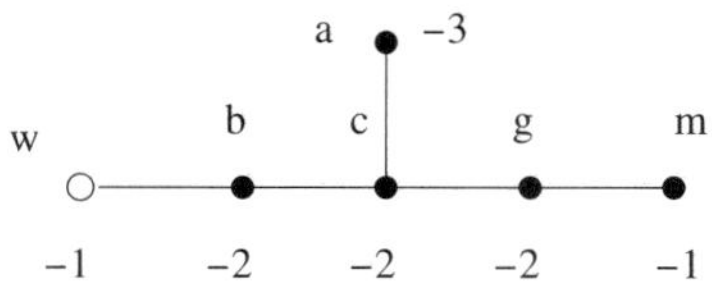

and finally let (S_0, p_0) be the parental system that consists of the common vertices of the two graphs above. The graph of (S_0, p_0) is

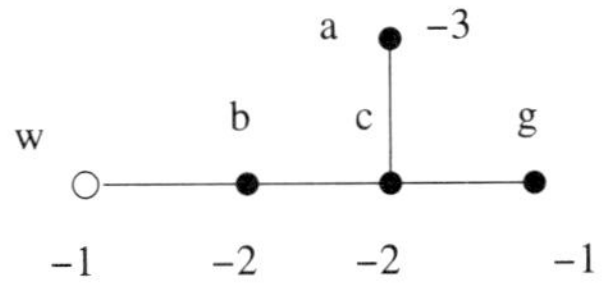

The composable triple $\mathcal{G}$, (S, p), (S_0, p_0) has two branches: a simply rooted branch $B_1 = \{g, h, i, j, k\}$, with root g, and a doubly rooted branch $B_2 = \{a, c, d, e, f\}$, with roots a and c. The set S_1 is the set $\{a, c, g, m\}$ and the (S_0, p_0)-composition of $\mathcal{G}$ and (S, p) is

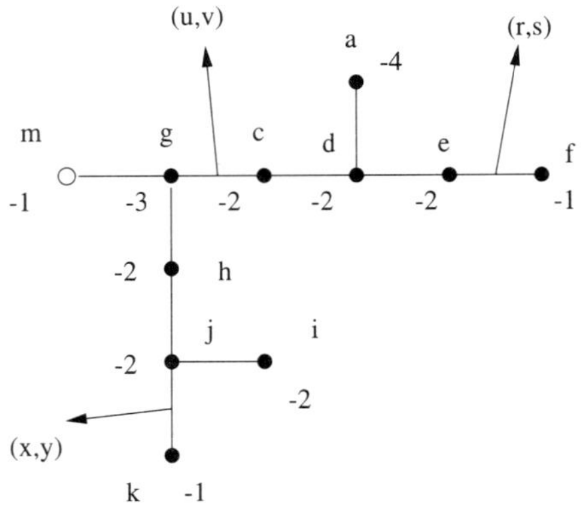

There are some conditions under which computing the (S_0, p_0)-composition of $\mathcal{G}$ and (S, p) for a composable triple is particularly easy:

REMARK 4.5. Let $\mathcal{G}$, (S, p), (S_0, p_0) be a composable triple such that there are two or more vertices of the first generation of $\mathcal{G}$; assume that we know that $\mathcal{G}_c$ satisfy the same property. Then S should be included in the set of vertices of $\mathcal{G}$ (or that is the same $S = S_0$), otherwise there would be exactly one vertex of the first generation in $\mathcal{G}_c$. In these conditions it is easy to check that $\mathcal{G}_c$ equals the result of letting the last vertex of S become the root of $\mathcal{G}$.

Now we apply our combinatorial construction to explain the change on the behaviour at infinity of a polynomial under composition with an automorphism: Let $f \in \mathbb{C}[x, y]$, let I be a partial resolution set, and let p_I be the parental function defined in I. Denote by $\mathcal{G}$ the decorated graph associated with f and I. Let $\phi \in Aut(\mathbb{C}^2)$ with graph $\mathcal{A}_{n_1,\ldots,n_r}$ for some numbers $r, n_1, \ldots, n_r$. Denote by S its associated set on infinitely near points, let p be the parental function defined in S. Define (S_0, p_0) by letting S_0 be the intersection $I \cap S$, and p_0 the restriction of p_I to S_0. It is easy to check that $\mathcal{G}$,(S, p),(S_0, p_0) is a composable triple.

DEFINITION 4.6. Let $f \in \mathbb{C}[x, y]$, let I be a partial resolution set and let $\phi \in Aut(\mathbb{C}^2)$. The composable triple $\mathcal{G}$,(S, p),(S_0, p_0) introduced in the last discussion is called *composable triple associated with* f, I, ϕ.

LEMMA 4.7. *Let* $f \in \mathbb{C}[x, y]$ *and let* $\mathcal{G}$ *be the graph of a partial resolution of it. Let* $\mathcal{G}$,(S, p),(S_0, p_0) *be a composable triple. Then there exist an automorphism* ϕ *such that* $\mathcal{G}$,(S, p),(S_0, p_0) *is the composable triple associated with* f, I, ϕ.

PROOF. As S_0 is identified with a subset of I, then each vertex of S_0 is an infinitely near point to $\mathbb{P}^2$. Let $\pi'' : Z \to \mathbb{P}^2$ be the composition of the blowing ups centered at the points of S_0. There is a set S' of infinitely near points to Z satisfying that no of them meets the strict transform of $\mathcal{C}$ by π'', and also that the set $S_0 \cup S'$, with the parental function obtained viewing it as a set of infinitely near points to $\mathbb{P}^2$, is a parental system isomorphic to (S,p). Let $\pi : X \to \mathbb{P}^2$ be the morphism obtained by composing the blowing ups centered at points of $S_0 \cup S'$ and let $\pi' : X \to \mathbb{P}_2$ the morphism obtained by successively contracting the all possible components of $\pi^* L$ starting by L. The restriction of $\pi' \circ \pi^{-1}$ to $\mathbb{C}^2$ is the automorphism that we need. □

NOTATION 4.8. Let $\psi \in G_{n_1,\dots,n_r}$; compactifying $\mathbb{C}^2$ to $\mathbb{P}^2$, the automorphism ψ can be extended to a birational morphism $\psi : \mathbb{P}^2 \to \mathbb{P}^2$. From now on we will denote the source (resp. target) $\mathbb{P}^2$ by $\mathbb{P}^2_s$ (resp. $\mathbb{P}^2_t$); we will denote the line at infinity of $\mathbb{P}^2_s$ (resp. $\mathbb{P}^2_t$) by L (resp L'). In the formula 2.1 from page 22 it was defined $\mathcal{I}$, a that certain subset of the set of infinitely near points to $\mathbb{P}^2$, we will denote this subset by $\mathcal{I}_{\mathbb{P}^2_s}$ (resp. $\mathcal{I}_{\mathbb{P}^2_t}$) when we want to denote infinitely near points to the source (resp. target) $\mathbb{P}^2$.

The automorphism ψ can be factorised as $\psi = \pi' \circ \pi^{-1}$ where $\pi : X \to \mathbb{P}^2_s$ and $\pi' : X \to \mathbb{P}^2_t$ are blowing up processes with graphs $\mathcal{A}$ and $\mathcal{A}'$ isomorphic to $\mathcal{A}_{n_1,\dots,n_r}$ and $\mathcal{A}_{n_r,\dots,n_1}$. Denote by (S,p) the parental system associated with the graph $\mathcal{A}$, and by (S,p') the parental system associated with the graph $\mathcal{A}'$.

LEMMA 4.9. *Let $f \in \mathbb{C}[x,y]$, let I be a partial resolution set and let $\psi \in Aut(\mathbb{C}^2)$. Let $\mathcal{G}$,(S,p),(S_0,p_0) be the composable triple associated with f, I, ψ and let $\mathcal{G}_c$ be the (S_0,p_0)-composition of $\mathcal{G}$ and (S,p). Then $\mathcal{G}_c$ is the graph of a partial resolution of $f \circ \psi^{-1}$.*

PROOF. It is easy to reduce to the case of $\psi \in G_{n_1,\dots,n_r}$ for certain $r, n_1, \dots, n_r$. We will use Notation 4.8. The set S viewed as the set of vertices of (S,p') is identified with the set of infinitely near points to $\mathbb{P}^2_t$ that are centers of the blowing ups whose composition is π'. Let m be the global minimum of the subset $S_0' \cup S_0'' \subset S$, needed for the construction of $\mathcal{G}_c$; let S_1 be as defined in formula 4.1; consider $B_1, \dots, B_s$ and $m_1, \dots, m_s$ as defined in definition 4.2. Each of the elements of any B_i is an element of I; consider the set

$$I_c := S_1 \bigcup \cup_{i=1}^{s} B_s \tag{4.2}$$

and define on it the parental function p_c by imposing that the restriction of p_c to any of the B_i's is p_I and that the restriction of p_c to S_1 is p'. As for each $i \leq s$ we have the inclusion $p(m_i) \subset S_1$, any of the elements $[x]$ of B_i is infinitely near to X. As there is a blowing up process $\pi' : X \to \mathbb{P}^2_t$, then $[x]$ can be viewed as an infinitely near point to $\mathbb{P}^2_t$; hence I_c can be viewed as a set of infinitely near points to $\mathbb{P}^2_t$ and p_c is just the natural parental function defined for sets of infinitely near points. With this we have proved that I_c is a *s.s.i.n.p.* to L'.

We need to show that I_c is a partial resolution set of $f \circ \psi^{-1}$, i.e. that $I_c \subset I_{f\circ\psi^{-1}}$. By the definition of the set $I_{f\circ\psi^{-1}}$, on formula 2.2 of page 22 we see that I_c must satisfy first, that $I_c \subset \mathcal{J}_{f\circ\psi^{-1}}$, and second, that $I_c \subset \mathcal{I}_{\mathbb{P}^2_t}$.

We check the first property. View the elements of S as infinitely near points to $\mathbb{P}^2_t$. Let $C \subset \mathbb{P}^2_s$ (resp. $C' \subset \mathbb{P}^2_t$) be the curve defined by f (resp. $f \circ \psi^{-1}$). Because of the definition of S_1 is clear that $S_1 \subset \mathcal{J}_{f\circ\psi^{-1}}$. We have to check that for any branch $\mathcal{B}_i$, the infinitely near points B_i are contained in $\mathcal{J}_{f\circ\psi^{-1}}$. We choose such a branch $\mathcal{B}_i$. Let $\pi'_1 : Z \to \mathbb{P}^2_t$ be the blowing up of $\mathbb{P}^2_t$ at the elements of S_1. Let $\pi'_2 : X \to Z$ be the blowing up of Z at the elements of $S \setminus S_1$. Then $\pi' = \pi'_1 \circ \pi'_2$. The variety X also is the result of blowing up the points of S but viewed as infinitely near points to $\mathbb{P}^2_s$. Such blowing up is $\pi : X \to \mathbb{P}^2_s$. Let $\tilde{C}$ be the strict transform by π of C. Note that $\tilde{C}$ meets m_i; moreover any point $x \in B_i$ is an infinitely near point meeting the strict transform of the germ $(\tilde{C}, m_i)$ by the minimal blowing up process $\sigma : Y \to X$ such that x has a representative in Y. Let $\tilde{C}'$ (resp. $\tilde{C}'_1$) denote the strict transform of C' by π' (resp. π'_1). The curves $\tilde{C}$ and $\tilde{C}'$ are equal, and hence, the germ $(\tilde{C}, m_i)$ is a germ of $\tilde{C}'$. As m_i is not a point of $S \setminus S_1$, we conclude that $(\tilde{C}, m_i)$ is also a germ of $\tilde{C}'_1$. Hence the set B_i viewed as a set of infinitely near points to $\mathbb{P}^2_t$ is also a subset of $\mathcal{J}_{f\circ\psi^{-1}}$.

To check the second property we remark that S_1 is obviously contained in $\mathcal{I}$. We need to verify that each B_i is contained in $\mathcal{I}_{\mathbb{P}^2_t}$. We start with some previous remarks: if B_i is simply rooted, the root w_i cannot be a free central vertex because then m_i would not belong to $\mathcal{I}$; for the same reason if B_i is doubly rooted then its root must consist of two central vertices. We consider several cases:

(1) If B_i is simply rooted and w_i is the fixed central vertex of the j-th floor. Let l be the lenght of the $j - th$ floor. As $B_i \subset \mathcal{I}_{\mathbb{P}^2_s}$, it follows that for any totally ordered chain $m_i =: x_1 \geq_{p_I} x_2 \geq_{p_I} ...$ of vertices of B_i the following properties apply: $p(x_1) = w_i$, $p(x_h) = x_{h-1}$ for $2 \leq h \leq l-1$; finally, $\{x_{l-1}, ..., x_k\}$, with the restriction of the parental function p_I to it, is isomorphic to a saturated subset of a parental set with graph $\mathcal{A}_{m_1,\dots,m_d}$ for some $d, m_1, ..., m_s$. View B_i and S as a sets of infinitely near points to $\mathbb{P}^2_t$. Notice that w_i is the fixed central vertex of the $r - j$-th floor of $\mathcal{A}'$, whose lenght is also l. Then, the parental system formed by the union of the vertices of $\mathcal{A}'$ bigger or equal that w_i and the vertices of the totally ordered chain given above is the parental system of an $\mathcal{A}$-constructible graph. This implies that $B_i \subset \mathcal{I}_{\mathbb{P}^2_t}$.

(2) If B_i is doubly rooted and $p(m_i) = \{w_{i,1}, w_{i,2}\}$ then $w_{i,1}$ and $w_{i,2}$ are the free and fixed central vertices of the $j - th$ floor of $\mathcal{A}$. Let l be the lenght of the $j - th$ floor. As $B_i \subset \mathcal{I}_{\mathbb{P}^2_s}$, it follows that for any totally ordered chain $m_i =: x_1 \geq_{p_I} x_2 \geq_{p_I} ...$ of vertices of B_i the following properties apply: $p(x_1) = \{w_{i,1}, w_{i,2}\}$, there is an l' such that if $2 \leq h \leq l'$ then $p(x_h) = \{w_{i,1}, x_{h-1}\}$, if $l'+1 \leq h \leq l+l'-1$ then $p(x_h) = x_{h-1}$; finally, $\{x_{l+l'-1}, ..., x_k\}$, with the restriction of the parental function p_I to it, is isomorphic to a saturated subset of a parental set with graph $\mathcal{A}_{m_1,\dots,m_d}$ for some $d, m_1, ..., m_s$. View B_i and S as a sets of infinitely near points to $\mathbb{P}^2_t$. Notice that $w_{i,1}$ and $w_{i,2}$ are the free and fixed central vertices of the $r - j$-th floor of $\mathcal{A}'$, which is also of lenght l. It is easy to check that the parental system formed by the union of the vertices of $\mathcal{A}'$ bigger or equal that w_i and the vertices of the totally ordered chain given above is the parental system of an $\mathcal{A}$-constructible graph. This implies that $B_i \subset \mathcal{I}_{\mathbb{P}^2_t}$.

(3) It remains to check the cases in which B_i is simply rooted and w_i is not the fixed central vertex. After reading the two cases above the reader should not have problems in cheking these cases by himself.

We have shown that I_c is a partial resolution set of $f\circ\psi$. To prove the lemma we only have to check that the weights and the decoration of $\mathcal{G}_c$ coincide with the weights and the decoration of the graph of the partial resolution of $f \circ \psi^{-1}$ associated with I_c. This is easy and we do not do it here. □

PROPOSITION 4.10. *Let $f \in \mathbb{C}[x, y]$ and $\phi \in Aut(\mathbb{C}^2)$. Let $\mathcal{B}_f$,(S,p),(S_0,p_0) be the composable triple associated with f,I_f,ϕ. If $\mathcal{B}_f$ denotes the behaviour graph of f then the behaviour graph of $f \circ \phi^{-1}$ is the (S_0,p_0)-composition of $\mathcal{B}_f$ and (S,p).*

PROOF. Denote by $\mathcal{B}'$ the (S_0,p_0)-composition of $\mathcal{B}_f$ and (S,p). From the last lemma we know that $\mathcal{B}'$ is a partial resolution graph of $f \circ \phi^{-1}$. As $\Lambda_2(\mathcal{B}_f)$ and $\Lambda_4(\mathcal{B}_f)$ are empty, by construction of $\mathcal{B}'$, we deduce that $\Lambda_2(\mathcal{B}')$ and $\Lambda_4(\mathcal{B}')$ are also empty. This implies that $\mathcal{B}'$ is the behaviour graph of $f \circ \phi^{-1}$. □

DEFINITION 4.11. Given $\mathcal{B}_1$, $\mathcal{B}_2$, behaviour graphs, we say that $\mathcal{B}_1 \sim \mathcal{B}_2$ if and only if there exist a composable triple $\mathcal{B}_1$, (S,p), (S_0,p_0) such that the (S_0,p_0) composition of $\mathcal{B}_1$ and (S,p) is $\mathcal{B}_2$.

It would be possible to prove combinatorially that $\sim$ is an equivalence relation, but we do not need to do it because, due to Lemma 4.7 and Proposition 4.10, the following remark is obvious, and it implies that $\sim$ is an equivalence relation.

REMARK 4.12. The following assertions are equivalent:

(1) $\mathcal{B}_1 \sim \mathcal{B}_2$.
(2) Each polynomial with behaviour graph $\mathcal{B}_1$ is R-equivalent to a polynomial with behaviour graph $\mathcal{B}_2$.
(3) There is a polynomial with behaviour graph $\mathcal{B}_1$ which is R-equivalent to a polynomial with behaviour graph $\mathcal{B}_2$.

Note that in the last remark we are using implicity that any behaviour graph is realizable by a polynomial.

4.2. Degree and composition by automorphisms

Let $f \in \mathbb{C}[x, y]$ and $\mathcal{G}$ a partial resolution graph of it. Proposition 2.49 implies that the graph $\mathcal{G}$ determines the degree of f. Here we give a formula of $\deg(f)$ in terms of $\mathcal{G}$. Define a function μ from the set of vertices of $\mathcal{G}$ to $\mathbb{N}$, the value $\mu(v)$ is defined inductively on the parental system: if w is the root then $\mu(w) = 1$, if $p(v) = \{v'\}$ then $\mu(v) = \mu(v')$ and if $p(v) := \{v', v''\}$ then $\mu(v) = \mu(v') + \mu(v'')$. If $\Lambda = \Lambda_1 \cup \Lambda_2 \cup \Lambda_3 \cup \Lambda_4$ is the decoration of $\mathcal{G}$, for each $a \in \Lambda$ define a function $\iota_a : \gamma(a) \to \mathbb{N}$ as follows: if $a \in \Lambda_1 \cup \Lambda_2$ and $\gamma(a) = \{v\}$, let $\iota_a(v)$ be the first component of the weight of a; if $a \in \Lambda_3 \cup \Lambda_4$ and $\gamma(a) = \{v, v'\}$ with $v \geq_p v'$, then let $\iota_a(v)$ be the first component of the weight of a and let $\iota_a(v')$ be the second component of the weight of a. Let π be the blowing up process that corresponds to the partial resolution with which we are working, each vertex v represents a component E_v of π^*L (where L is the line at infinity), and $\mu(v)$ is the multiplicity of E_v in the divisor π^*L. Given an arrow a, it represents one or several branches of the strict transform of C by π, that intersect π^*L at E_v; the number $\iota_a(v)$ is the sum of the intersection multiplicities of these branches with E_v.

With this terminology, using Bezout's theorem and the formulae that explain the behaviour of intersection multiplicity under blowing up we get:

$$\deg(f) = \sum_{a \in \Lambda} \sum_{v \in \gamma(a)} \mu(v)\iota_a(v) \tag{4.3}$$

The function μ does not depend on the decoration and its definition makes sense for any parental graph. Now we describe the function μ for the graph $\mathcal{A} := \mathcal{A}_{n_1,..,n_r}$, for any $s, n_1, ..., n_r$: if the i-th floor consists of the vertices $\{v_0, ..., v_{2n_i-1}\}$, ordered by the parental ordering, then

$$(4.4)\qquad \begin{array}{rclll} \mu(v_0) & = & \mu(v_1) & = & n_1...n_{i-1} \\ \mu(v_j) & = & n_1...n_{i-1}j & \text{for} & 2 \leq j \leq n_i \\ \mu(v_j) & = & n_1...n_{i-1}n_i & \text{for} & n_i \leq j \leq 2n_i - 1 \end{array}$$

Let $\mathcal{A}'$ be the graph obtained from $\mathcal{A}$ by letting the last vertex become the root. The graphs $\mathcal{A}$ and $\mathcal{A}'$ have the same set of vertices S. We denote by μ' the function analogous to μ but considering in S the parental ordering coming from $\mathcal{A}'$. The values of μ' at the vertices of the i-th floor of $\mathcal{A}$, that is the $(r-i)$-th floor of $\mathcal{A}'$ are:

$$(4.5)\qquad \begin{array}{rcll} \mu'(v_0) & = & n_i n_{i+1}...n_r & \\ \mu'(v_1) & = & n_{i+1}...n_r & \\ \mu'(v_j) & = & (2n_i - j)n_{i+1}...n_i & \text{for} \quad n_i \leq j \leq 2n_i - 1 \\ \mu'(v_j) & = & n_i n_{i+1}...n_r & \text{for} \quad 2 \leq j \leq n_i \end{array}$$

Consider a polynomial f and automorphism $\psi \in G$. Extend ψ to a birational transformation of $\mathbb{P}^2$ and factorize it as $\psi = \pi' \circ \pi^{-1}$, where $\pi : X \to \mathbb{P}^2_s$ and $\pi' : X \to \mathbb{P}^2_t$ are blowing up processes. Assume that the graph of π (resp. π') is $\mathcal{A}$ (resp $\mathcal{A}'$). Let $\tilde{C}$ be the strict transform of C by π. Although π is not necessarily a partial resolution of f, we can use $\tilde{C}$ to add a decoration to $\mathcal{A}$ in the same way that we did for partial resolutions in Construction 2.44. Denote by C' the curve defined in $\mathbb{P}^2_t$ by $f \circ \psi^{-1}$. The strict transform of C' by π' is also $\tilde{C}$. Use it to add a decoration to $\mathcal{A}'$. The decorated graphs $\mathcal{A}$ and $\mathcal{A}'$ are identical except in that the root of $\mathcal{A}$ is the last vertex of $\mathcal{A}'$ and viceversa; hence, for any arrow a, the function ι_a is the same for both of them. Denote by S the common set of vertices of both graphs. Because of formula 4.3 we get the following formulae that help us in comparing the degrees of f and $f \circ \psi^{-1}$:

$$(4.6)\qquad \begin{array}{rcl} \deg(f) & = & \sum_{v \in I} \mu(v)\iota(v) \\ \deg(f \circ \psi^{-1}) & = & \sum_{v \in I} \mu'(v)\iota(v) \end{array}$$

If we take instead of $\psi \in G$ a general automorphism $\phi \in Aut(\mathbb{C}^2)$ then we have a factorisation $\phi = \theta \circ H \circ \psi$. As $\deg(f \circ \phi^{-1}) = \deg(f \circ \psi^{-1})$ the formulae 4.6 allow also to compare $\deg(f \circ \phi^{-1})$ and $\deg(f)$.

The following theorem is due to P.G.Wightwick, see [**28**], page 2. Using the formulae 4.6 we are going to give a proof of it within the techniques of this memoir.

THEOREM 4.13. (P.G.Wightwick.) *Let $f \in \mathbb{C}[x, y]$ and $\phi \in Aut(\mathbb{C}^2)$ such that the decomposition of ϕ in primitive automorphisms is $\phi = \theta \circ H \circ \psi_r \circ ... \circ \psi_1$ for $r \geq 2$. Let $f_i := f \circ (\psi_i \circ ... \circ \psi_1)^{-1}$ for any $1 \leq i \leq r$ and $f_0 := f$. Then sequence of degrees $\deg(f_i)$ satisfies*

$$\deg(f_0) \geq ... \geq \deg(f_k) \leq ... \leq \deg(f_r).$$

Moreover there is at most one i such that $\deg(f_i) = \deg(f_{i+1})$ and, if this happens, then $\deg(f_i)$ is the minimal degree of the sequence.

PROOF. The proof is trivial after the proof of the following lemma. □

LEMMA 4.14. *Let $f \in \mathbb{C}[x, y]$ and $\psi \in G$ such that $\deg(f \circ \psi^{-1}) \leq \deg(f)$. Let $\psi = \psi_r \circ ... \circ \psi_1$ be the decomposition of ψ in primitive automorphisms. Then $\deg(f \circ \psi_1^{-1}) \leq \deg(f)$. Moreover, the inequality is strict if $\deg(f \circ \psi^{-1}) < \deg(f)$ or if $r > 1$.*

PROOF. The graph of ψ is $\mathcal{A}_{n_1,...,n_r}$ for some numbers $n_1, .., n_r$. Decompose ψ as $\psi = \pi' \circ \pi^{-1}$, and denote by $\tilde{C}$ the strict transform of C by π. Use $\tilde{C}$ to give a decoration Λ to $\mathcal{A}_{n_1,...,n_r}$. Let the functions μ, μ', and, for each arrow a, the function ι_a, be defined as in page 67. Let A_1 be the set of arrows such that its set of attaching vertices is included in the first floor of $\mathcal{A}_{n_1,...,n_r}$, let $A_2 := \Lambda \setminus A_1$. Define the following numbers:

$$\alpha := \sum_{a \in A_1} \sum_{v \in \gamma(a)} \mu(v) \iota_a(v) \quad \beta := \sum_{a \in A_2} \sum_{v \in \gamma(a)} \mu(v) \iota_a(v)$$

$$\alpha' := \sum_{a \in A_1} \sum_{v \in \gamma(a)} \mu'(v) \iota_a(v) \quad \beta' := \sum_{a \in A_2} \sum_{v \in \gamma(a)} \mu'(v) \iota_a(v)$$

Because of formulae 4.6 we have that $\deg(f) = \alpha + \beta$ and $\deg(f \circ \psi^{-1}) = \alpha' + \beta'$.

The graph of ψ_1 is $\mathcal{A}_{n_1}$ and its set of vertices is the set of vertices of the first floor of $\mathcal{A}_{n_1,...,n_r}$. Decompose $\psi_1 = \pi_1' \circ \pi_1^{-1}$ and denote by $\tilde{C}_1$ the strict transform of C by π_1. Use it to give a decoration Λ' to $\mathcal{A}_1$. Let μ_1, μ_1' and $\iota_{1,a}$ (for any arrow $a \in \Lambda'$), be defined for $\mathcal{A}_{n_1}$ as in page 67. As the arrows of A_1 correspond to intersection points of $\tilde{C}_1$ and $\pi_1^* L$ that are not centers of any blowing up of π, for each of these arrows there is an arrow in Λ' with the same weight; denote by A_1' the arrows of Λ' that correspond to arrows of A_1. The set $A_2' := \Lambda' \setminus A_1'$ has a unique arrow b that is attached to the last vertex of $\mathcal{A}_{n_1}$; we denote this last vertex by v'. Define

$$\alpha_1 := \sum_{a \in A_1'} \sum_{v \in \gamma(a)} \mu_1(v) \iota_a(v) \quad \beta_1 := \mu_1(v') \iota_{1,b}(v')$$

$$\alpha_1' := \sum_{a \in A_1'} \sum_{v \in \gamma(a)} \mu_1'(v) \iota_a(v) \quad \beta_1' := \mu_1'(v') \iota_{1,b}(v')$$

Because of formulae 4.6, $\deg(f) = \alpha_1 + \beta_1 =$ and $\deg(f \circ \psi_1^{-1}) = \alpha_1' + \beta_1'$.

For any vertex v in the first floor and any arrow $a \in A_1$ we have $\mu(v) = \mu_1(v)$, $\mu'(v) = n_2 ... n_r \mu_1'(v)$ and $\iota_a(v) = \iota_{1,a}(v)$. Hence $\alpha = \alpha_1$ and $\alpha' = n_2 ... n_r \alpha_1'$. If A_2 is empty then $\beta = \beta' = \beta_1 = \beta_1' = 0$ and the lemma is obvious in this case. Assume that A_2 is not empty (in this case $r \geq 2$). As $\deg(f) = \alpha + \beta = \alpha_1 + \beta_1$, hence $\beta = \beta_1$. As $\mu_1(v') = n_1$ and $\mu_1'(v') = 1$ then $\beta_1 = n_1 \beta_1'$. Let

$$\beta'' := \frac{\beta_1'}{n_2 ... n_r}$$

We have the strict inequality $\beta' < n_2 ... n_r \beta_1'$: in fact, the closer are the arrows of A_2 to v' the bigger is β'; the limit case would be the case in which all these arrows were attached to v' and then β' would be $n_2 ... n_r \beta_1'$, because $\mu'(v') = n_2 ... n_r$; this limit case is impossible because the arrows of A_2 are not attached to the first floor of $\mathcal{A}_{n_1,...,n_r}$. We claim the inequality $\beta' \geq \beta''$: in fact, the closer are the arrows of A_2 to the last vertex of $\mathcal{A}_{n_1,...,n_r}$ the bigger is β_1'/β', the limit case would be the case in which all these arrows were attached to the last vertex, and then it is easy to check that then β'_1/β' would be $n_2...n_r$. By assumption $\deg(f) \geq \deg(f\circ\psi)$; the statements above imply

$$\alpha + \beta = \alpha + n_1...n_r\beta'' \geq \alpha' + \beta' = n_2...n_r\alpha'_1 + \beta' \geq n_2...n_r\alpha'_1 + \beta'' \tag{4.7}$$

hence

$$\alpha'_1 \leq \frac{\alpha + (n_1...n_r - 1)\beta''}{n_2...n_r}. \tag{4.8}$$

We want to prove $\deg(f) > \deg(f\circ\psi_1^{-1})$ or, what is equivalent

$$\alpha_1 + \beta_1 = \alpha + n_1...n_r\beta'' > \alpha'_1 + \beta'_1 = \alpha'_1 + n_2...n_r\beta'' \iff \tag{4.9}$$

$$\alpha'_1 < \alpha + (n_1 - 1)n_2...n_r\beta'' \tag{4.10}$$

Subtituting 4.8 in 4.10 we are reduced to prove

$$\frac{\alpha + (n_1...n_r - 1)\beta''}{n_2...n_r} < \alpha + (n_1 - 1)n_2...n_r\beta'' \iff \tag{4.11}$$

$$\alpha(n_2...n_r - 1) + \beta''\big((n_1 - 1)(n_2...n_r)^2 - n_1...n_r + 1\big) > 0. \tag{4.12}$$

If $\alpha > 0$ then $\alpha(n_2...n_r - 1) > 0$, if $\alpha = 0$ the statement of the lemma is trivial. On the other hand

$$\begin{aligned} (n_1 - 1)(n_2...n_r)^2 - n_1...n_r + 1 &= \\ (n_1 - 1)(n_2...n_r - 1)^2 + (n_1 - 2)(n_2...n_r - 1) &> 0 \end{aligned}$$

This proves the lemma. □

Neumann and Wightwick, in their paper [**28**], state that if $f \in \mathbb{C}[x, y]$, and $\phi \in Aut(\mathbb{C}^2)_{n_1,...,n_r}$ then $r \leq 2(\deg(f) + \deg(f\circ\phi^{-1}))$. From the statement of their theorem the following sharper bound is possible: let

$$d(f) := \text{minimal degree of } f \text{ in its R-equivalence class}$$

$$\text{diff}(f) := \deg(f) - d(f)$$

COROLLARY 4.15. *If $f \in \mathbb{C}[x, y]$ and $\phi \in Aut(\mathbb{C}^2)_{n_1,...,n_r}$ then $r \leq \text{diff}(f) + \text{diff}(f\circ\phi^{-1}) + 1$. In particular if both f and $f\circ\phi^{-1}$ are of minimal degree then $r = 1$.*

This bound is achieved: the most obvious example is $f = x - y^2$ and $\phi = \phi_1\circ\phi_2\circ\phi_1$ where $\phi_1(x, y) = (x - y^2, y)$ and $\phi_2(x, y) = (x, y + x^2)$. They also state that if $f \in \mathbb{C}[x, y]$ and ψ is primitive then $\deg(\psi) \leq \max(\deg(f), \deg(f\circ\psi))$. This is almost always true, the only exception being the case in which f is a polynomial without monomials involving the variable y and ϕ a triangular automorphism. We make this precise in the following

PROPOSITION 4.16. *Let $f \in \mathbb{C}[x, y]$ and ψ a primitive automorphism. If f does not define a set of parallel (may be non reduced) lines then*

$$\deg(\psi) \leq \max(\deg(f), \deg(f\circ\psi)^{-1})$$

PROOF. Decompose ψ into $\pi'\circ\pi^{-1}$, let $\tilde{C}$ be the strict transform of C by π and let $\mathcal{A}_n$ be the graph of ψ. Add the decoration corresponding to $\tilde{C}$ to $\mathcal{A}_n$. If there is an arrow such that one of its attaching points is a vertex v different from the free central vertex, then, either $\mu(v)$ or $\mu'(v)$ equals n. Hence, either $\deg(f) \geq n$ or $\deg(f\circ\psi^{-1}) \geq n$. The case in which the decoration has a unique arrow attached to the free central vertex is the case in which f has a unique point at infinity and all the branches through it are transversal to the line at infinity, then f defines a set of parallel lines. □

The following proposition can be easily proved after the results above:

PROPOSITION 4.17. *Let $f \in \mathbb{C}[x, y]$ and $\phi \in Aut(\mathbb{C}^2)$. If f is not R-equivalent to a polynomial that defines a set of parallel (may be non reduced) lines then*

$$\deg(\phi) \leq \frac{\deg(f)!\deg(f \circ \phi^{-1})!}{d(f)!(d(f) - 1)!}$$

4.3. Invariance of the behaviour graph by translations on the left

The purpouse of this section is to answer the following question of D. Siersma: given $f \in \mathbb{C}[x, y]$ and $t \in \mathbb{C}$, can the behaviour graphs of f and $f + t$ be different?

Consider $f \in \mathbb{C}[x, y]$ with behaviour graph $\mathcal{B}_f$, let v be a vertex of $\mathcal{B}$. Let $\mathcal{B}_{\geq v}(f)$ be the graph obtained by deletion of all the vertices that are not bigger than or equal to v by the parental ordering. To each arrow a of $\Lambda_2(\mathcal{B}_{\geq v}(f)) \cup \Lambda_4(\mathcal{B}_{\geq v}(f))$ we associate the point of I_f that is the infinitely near point corresponding to the vertex of $\mathcal{B}$ whose deletion gives rise to a. We define $I_{\geq v}(f)$ to be the subset of I_f consisting of the infinitely near points associated with the vertices of $\mathcal{B}_{\geq v}(f)$ and to the arrows of $\Lambda_2(\mathcal{B}_{\geq v}(f)) \cup \Lambda_4(\mathcal{B}_{\geq v}(f))$.

LEMMA 4.18. *Consider any $f \in \mathbb{C}[x, y]$ and $t \in \mathbb{C}$. Let $v \in \mathcal{B}_f$ be a vertex of the first floor, then there exist a vertex $v(t) \in \mathcal{B}_{f+t}$ such that $\mathcal{B}_{\geq v}(f)$ and $\mathcal{B}_{\geq v(t)}(f + t)$ are isomorphic and moreover the sets $I_{\geq v}(f)$ and $I_{\geq v(t)}(f + t)$ are equal.*

PROOF. We show the lemma by induction on the parental ordering of $\mathcal{B}$. Assume first that v is the root. Then $I_{\geq v}(f)$ is the set of points at infinity of the curve $f = 0$, and $\mathcal{B}_{\geq v}(f)$ consists of the root together with an arrow of $\Lambda_2(\mathcal{B}_{\geq v}(f))$ for each point of $I_{\geq v}(f)$. We have to choose $v(t)$ to be the root of $\mathcal{B}_{f+t}$. The location of the points at infinity of $f + t = 0$ does not depend on t, hence $I_{\geq v(t)}(f + t)$ is independent on t; let $I_{\geq v(t)}(f + t) = \{x_1, ..., x_k\}$ for any t. By definition of partial resolution graphs we find an arrow $a_i(t) \in \Lambda_2(\mathcal{B}_{\geq v(t)}(f + t))$ associated to each x_i. If $x_i = (0 : 1 : 0)$ an easy argument using the Newton polygon of f shows that the weight of $a_i(t)$ does not depend on t. If $x_i \neq (0 : 1 : 0)$ we consider a linear automorphism H such that $H(0 : 1 : 0) = x_i$; note that there is a natural isomorphism between $\mathcal{B}_{f+t}$ and $\mathcal{B}_{(f+t) \circ H} = \mathcal{B}_{f \circ H + t}$; this gives rise an isomorphism between $\mathcal{B}_{\geq v(t)}(f + t)$ and $\mathcal{B}_{\geq v(t)}(f \circ H + t)$ that takes the arrow associated with x_i of the first graph to the arrow associated with $(0 : 1 : 0)$ of the second. Therefore we are reduces to the case of $x_i \neq (0 : 1 : 0)$. With this we have shown the lemma in the case of v being the root.

Now we deal with the case of v being either a central or a fixed vertex of $\mathcal{B}_f$. Let $\{w, v_1, ..., v_r = v\}$ be the set of vertices of $\mathcal{B}_{\geq v}(f)$; denote the arrows of $\Lambda_2(\mathcal{B}_{\geq v}(f))$ by a^i_j where $i \in \{0, 2, ..., r\}$, $1 \leq j \leq s_i$ for certain $s_i \geq 0$, and $\gamma_2(a^i_j) = v_i$ (note that when $s_i = 0$ there is no a^i_j). Then the subset of free points of $I_{\geq v}(f)$ consists of the point $x_1 \in L$ corresponding to the vertex v_1, together with a point x^i_j associated with each arrow a^i_j. Composing with a linear automorphism H we can assume that $x_1 = (0 : 1 : 0)$. The points $x_1, x^0_1, ..., x^0_{s_0}$ are the points at infinity of f, and hence the points at infinity of $f + t$ for any $t \in \mathbb{C}$; this ensures the presence of a vertex $v_1(t)$ in $\mathcal{B}_{f+t}$ whose associated infinitely near point is x_1.

If $r = 1$ we let $v(t) = v_1(t)$. In this case $I_{\geq v(t)}(f+t)$ consists of the points at infinity of $f+t$, this shows the independence of $I_{\geq v(t)}(f+t)$ of t. Any arrow of $\mathcal{B}_{\geq v(t)}(f+t)$ is either a_j^0 for a certain j, or an arrow attached to $v(t)$ or to the only edge that meets it. By induction we know that the weights of the a_j^0's do not depend on t. Using that $x_1 = (0:1:0)$ and an easy argument using the Newton polygon of f, we see that the weights of the arrows attached either to $v(t)$ or to the edge that meets it are also independent on t. This shows that $\mathcal{B}_{\geq v(t)}(f+t)$ does not depend on t.

Consider now $r > 1$. The presence of $v_r \in \mathcal{B}_f$ implies that there is an arrow a attached to the only edge that meets v_{r-1} in $\mathcal{B}_{\geq v_{r-1}}(f)$. By induction there exists $v_{r-1}(t) \in \mathcal{B}_{f+t}$ such that $I_{\geq v_{r-1}(t)}(f+t)$ does not depend on t and $\mathcal{B}_{\geq v_{r-1}}(f)$ is isomorphic to $\mathcal{B}_{\geq v_{r-1}(t)}(f+t)$. Thus, for any $t \in \mathbb{C}$, there exists an arrow $a(t)$ attached to the only edge that meets $v_{r-1}(t)$ in $\mathcal{B}_{\geq v_{r-1}(t)}(f+t)$; this, in turn, implies the existence of a vertex $v_r(t) \in \mathcal{B}_{f+t}$ such that $p(v_r(t)) = \{v_1(t), v_{r-1}(t)\}$ (where $w(t), v_1(t), ..., v_{r-1}(t)$ denote the vertices of $\mathcal{B}_{\geq v_{r-1}(t)}(f+t)$). We take $v(t) := v_r(t)$. With this choice the parental systems of $\mathcal{B}_{\geq v_r(t)}(f+t)$ and $\mathcal{B}_{\geq v_r}(f)$ are isomorphic. Denote the arrows of $\Lambda_2(\mathcal{B}_{\geq v_r(t)}(f+t))$ by $a_j^i(t)$ where $i \in \{0, 2, ..., r\}$, $1 \leq j \leq s_i(t)\}$ for certain $s_i(t) \geq 0$, and $\gamma_2(a_j^i(t)) = v_i(t)$; the subset of free points of $I_{\geq v_r(t)}(f+t)$ consists of the point $x_1(t) \in L$ corresponding to the vertex $v_1(t)$, together with a point $x_j^i(t)$ associated with each arrow $a_j^i(t)$.

By induction we deduce that if $i < r$ neither $s_i(t)$, nor $x_j^i(t)$ for any $j \leq s_j^i(t)$, nor the weight of any arrow of $\mathcal{B}_{\geq v_{r-1}(t)}(f+t)$ depends on t. The only possible arrows of $\mathcal{B}_{\geq v_r(t)}(f+t)$ that are not already present in $\mathcal{B}_{\geq v_{r-1}(t)}(f+t)$ are the $a_j^r(t)$'s, and two arrows $b(t) \in \Lambda_3(\mathcal{B}_{\geq v_r(t)}(f+t))$ and $c(t) \in \Lambda_4(\mathcal{B}_{\geq v_r(t)}(f+t))$, satisfying $\gamma_3(b(t)) = \{v_{r-1}(t), v_r(t)\}$ and $\gamma_4(c(t)) = \{v_1(t), v_r(t)\}$. The weight of $a(t)$ determines the weight of $b(t)$, the primary components of the weight of $c(t)$ and the sum of the primary components of the weights of the $a_j^r(t)$'s; hence these numbers do not depend on t.

Let $\pi : X \to \mathbb{P}^2$ be the composition of the blowing ups corresponding to the vertices $v_1(t), ..., v_r(t)$ (notice that π does not depend on t for being determined by the position of $x_1(t) = (0:1:0)$ and the parental function), denote by E_r the exceptional divisor corresponding to $v_r(t)$ and by $\tilde{C}(t)$ the strict transform of the curve $f+t=0$ by π; then

$$\{x_j^r(t) : 1 \leq j \leq s_j(t)\} = \dot{E}_r \cap \tilde{C}(t)$$

and the primary component $w_j(t)$ of the arrow $a_j^r(t)$ is $w_j(t) = I_{x_j^r(t)}(E, \tilde{C}(t))$. It follows that, if there is a value t_0 such that the sequence $w_1(t_0), ..., w_{s_r(t_0)}(t_0)$ differs from the sequence $w_1(t), ..., w_{s_r(t)}(t)$ for generic t, then at least two points among the $x_j^r(t)$'s tend to collide when t approaches t_0. Suppose that this happens, then for certain t there exists $l \leq s_r(t)$ such that $x_l^r(t)$ is different to any $x_j^r(t_0)$ for $j \leq s_r(t_0)$. Consider a primitive automorphism ϕ with graph $\mathcal{A}_r$ such that $x_l^r(t)$ is one of the blowing up centers of the minimal resolution of its indeterminancy. Using the formulae 4.6 it is easy to conclude that $\deg((f+t)\circ\phi^{-1}) < \deg((f+t_0)\circ\phi^{-1})$, which is impossible because $(f+t)\circ\phi^{-1} = f\circ\phi^{-1}+t$ and hence its degree equals $\deg(f\circ\phi^{-1})$. This shows that $s_r(t)$, $x_j^r(t)$ and $w_j(t)$ for any $j \leq s_r(t)$ are independent of t.

For any t and any $j \leq s_j(t)$ the existence of the arrow $a_j^r(t)$ implies the existence of a vertex $v_j^r(t) \in \mathcal{B}_{f+t}$ such that $p(v_j^r(t)) = v_r(t)$ and that the infinitely near point associated with $v_j^r(t)$ is $x_j^r(t)$. The possible arrows of $\mathcal{B}_{\geq v_j^r(t)}(f+t)$ that were not present in $\mathcal{B}_{\geq v_r(t)}(f+t)$ are an arrow $d(t) \in \Lambda_3(\mathcal{B}_{\geq v_j^r(t)}(f+t))$ such that $\gamma_3(d(t)) = \{v_r(t), v_j^r(t)\}$ and some arrows $e_1(t), ..., e_{k(t)}(t)$ of $\Lambda_2(\mathcal{B}_{\geq v_j^r(t)}(f+t))$ such that $\gamma_2(e_i(t)) = v_j^r(t)$ for any $i \leq k(t)$; let $(\alpha_1(t), \alpha_2(t))$ be the weight of $d(t)$ and, for $i \leq k(t)$, let $\beta_i(t)$ be the primary component of the weight $e_i(t)$. Let $m_j(t)$ and $n_j(t)$ denote respectively the multiplicity component and the secondary component of the weight of $a_j^r(t)$. If $m_j(t)$ is not independent of t then $m_j(t_0) > m_j(t)$ for t generic and certain special t_0. The formulae that define how to construct the weights of the arrows in the procedure of deletion of vertices of decorated graphs imply that

$$(4.13) \qquad w_j(t) = m_j(t) + \alpha_1(t) \quad \alpha_2(t) = n_j(t) \quad m_j(t) = \alpha_2(t) + \sum_{i=1}^{k(t)} \beta_i(t)$$

Consider a primitive automorphism ϕ with graph $\mathcal{A}_r$ such that, if I_ϕ is the set of blowing up centers of the minimal resolution of its indetermination, then the sets $I_\phi \cap I_{f+t}$ and $I_\phi \cap I_{f+t_0}$ are equal and consist of the infinitely near points bigger than or equal to $x_j^r(t) = x_j^r(t_0)$ in the parental ordering. Then using the formulae 4.6 we obtain

$$\deg((f+t)\circ\phi^{-1}) - \deg((f+t_0)\circ\phi^{-1}) =$$
$$r\alpha_1(t) + (r-1)(\alpha_2(t) + \sum_{i=1}^{k(t)} \beta_i(t)) - r\alpha_1(t_0) - (r-1)(\alpha_2(t_0) + \sum_{i=1}^{k(t_0)} \beta_i(t_0))$$

Substituting the equalities 4.13 we get

$$\deg((f+t)\circ\phi^{-1}) - \deg((f+t_0)\circ\phi^{-1}) =$$
$$r(w_j(t) - m_j(t)) + (r-1)m_j(t) - r(w_j(t_0) - m_j(t_0)) - (r-1)m(t_0) =$$
$$m_j(t_0) - m_j(t)) > 0$$

The last inequality is impossible, as we have seen before, hence we deduce that $m_j(t)$ is independent of t for any $j \leq s_r(t)$.

Now we are going to prove that $n_j(t)$ is independent of t for any $j \leq s_r(t)$. Fix $j \leq s_r(t)$. For any $h \leq k(t)$ there is an infinitely near point $y_h(t) \in I_{\geq v_j^r(t)}(f+t)$ corresponding to the arrow e_h. Let $\pi : X \to \mathbb{P}^2$ be the composition of the blowing ups corresponding to the vertices $v_1(t), ..., v_r(t), v_j^r(t)$; let E_j^r be the exceptional divisor corresponding to $v_j^r(t)$ and $\tilde{C}(t)$ the strict transform of the curve $f + t = 0$ by π. Then $n_j(t) = I_{E_r \cap E_j^r}(\tilde{C}(t), E_j^r)$,

$$\{y_h(t) : 1 \leq h \leq k\} = \dot{E}_j^r \cap \tilde{C}(t),$$

and $\beta_h(t) = I_{y_h(t)}(\tilde{C}(t), E_j^r)$. Due to the independence on t of $m_j(t)$, and to the third equation of 4.13, if there is a value t_0 such that $n_j(t_0) > n_j(t)$ for t generic, then, when t approaches t_0, at least one $y_h(t)$ should tend to $E_r \cap E_j^r$, and hence move its position when t moves. If $r = 2$ we consider the primitive automorphism ϕ with graph $\mathcal{A}_2$, and such that $x_j^r(t)$ is among the centers of its resolution of indeterminancy. Then the $y_h(t)$'s can be viewed as points at infinity of $(f+t)\circ\phi^{-1} = f\circ\phi^{-1} + t$, and hence they do not move when t moves; we get a contradiction. If $r > 2$, for a certain t there is a $y_h(t)$ which is different to all the

$y_h(t_0)$'s. Consider a primitive automorphism ϕ with graph $\mathcal{A}_r$ such that $y_h(t)$ is one of the blowing up centers of the minimal resolution of its indeterminancy. Using the formulae 4.6 it is easy to conclude that $\deg((f+t)\circ\phi^{-1}) < \deg((f+t_0)\circ\phi^{-1})$, which gives a contradiction. This shows that $n_j(t)$ is independent of t. With this we have shown that the weight of the arrow $a_j^r(t)$ is independent of t for any $j \leq s_r(t)$.

To finish the proof of the induction step for v_r we only need to show that the weight of the arrow $c(t)$ is independent of t. The primary components of this weight are determined by the weight of $a(t)$, and hence are independent of t. The proof that the secondary components are independent of t is analogous to the proof that the secondary component of the $a_j^r(t)$'s are independent of t.

It only remains to prove the induction step for free lateral vertices. As it is quite similar to the case of fixed vertices we leave it to the reader. □

THEOREM 4.19. *Given any $f \in \mathbb{C}[x,y]$, for any $t \in \mathbb{C}$, we have that $\mathcal{B}_f \simeq \mathcal{B}_{f+t}$ and $I_f = I_{f+t}$*

PROOF. Consider $f \in \mathbb{C}[x,y]$ with behaviour graph $\mathcal{B}_f$ and let (I_f, p) be the parental system associated with it. Given any parental subsystem $(S,p) \subset (I_f,p)$ we consider the graph $\mathcal{B}_S(f)$ obtained by deletion of all the vertices that are not in S. We associate with each arrow a of $\Lambda_2(\mathcal{B}_S(f)) \cup \Lambda_4(\mathcal{B}_S(f))$ the point of I_f that corresponds to the vertex of $\mathcal{B}$ whose deletion gives rise to a. We define $I_S(f)$ to be the subset of I_f consisting of the infinitely near points associated with the vertices of $\mathcal{B}_S$ and to the arrows of $\Lambda_2(\mathcal{B}_S(f)) \cup \Lambda_4(\mathcal{B}_S(f))$.

Let (S,p) be the maximal parental subsystem of (I_f,p) such that, for any t, there exists a parental subsytem $(S(t),p)$ of (I_{f+t},p) satisfying that the graphs $\mathcal{B}_S(f)$ and $\mathcal{B}_{S(t)}(f+t)$ are isomorphic, and that $I_S(f) = I_{S(t)}(f+t)$. If $(S,p) = (I_f,p)$ then the theorem is true. Assume the contrary; then there exist a vertex $v \in \mathcal{B}_f$ satisfying $p(v) \subset S$ and a certain t such that there is no vertex $v(t) \in \mathcal{B}_{f+t}$ with the following properties:

- $p(v(t)) \subset S(t)$ and $\mathcal{B}_{S\cup\{v\}}(f)$ is isomorphic to $\mathcal{B}_{S(t)\cup\{v(t)\}}(f+t)$.
- $I_{S\cup\{v\}}(f){=}I_{S(t)\cup\{v(t)\}}(f+t)$.

The deletion of the vertex v of $\mathcal{B}_{S\cup\{v\}}(f)$ gives rise to an arrow $a \in \Lambda_2(\mathcal{B}_S(f)) \cup \Lambda_4(\mathcal{B}_S(f))$, which has associated an infinitely near point $x \in I_S(f)$. As $\mathcal{B}_S(f) \cong \mathcal{B}_{S(t)}(f+t)$ and $I_S(f) = I_{S(t)}(f+t)$ for any $t \in \mathbb{C}$, there exists an arrow $a(t) \in \Lambda_2(\mathcal{B}_{S(t)}(f+t)) \cup \Lambda_4(\mathcal{B}_{S(t)}(f+t))$ with the same weight than a whose associated infinitely near point is also x. This implies the existence of a vertex $v(t) \in \mathcal{B}_{f+t}$ whose deletion gives rise to $a(t)$ and whose associated infinitely near point is x. It is clear that the parental systems $(S \cup \{v\}, p)$ and $(S(t) \cup \{v(t)\}, p)$ are isomorphic. Moreover:

- Suppose that $\mathcal{B}_{S\cup\{v\}}$ and $\mathcal{B}_{S(t)\cup\{v(t)\}}$ are not isomorphic. As $(S\cup\{v\},p) \cong (S(t) \cup \{v(t)\}, p)$, the graphs obtained by deletion of decoration from $\mathcal{B}_{S\cup\{v\}}$ and $\mathcal{B}_{S(t)\cup\{v(t)\}}$ are be isomorphic. Therefore the difference between $\mathcal{B}_{S\cup\{v\}}$ and $\mathcal{B}_{S(t)\cup\{v(t)\}}$ must be in the decoration. Consider the decoration Λ (resp. $\Lambda(t)$) of $\mathcal{B}_{S\cup\{v\}}$ (resp. $\mathcal{B}_{S(t)\cup\{v(t)\}}$); define the subset $A_1 \subset \Lambda$ (resp. $A_1(t) \subset \Lambda(t)$ to be the set of arrows attached either to v (resp. $v(t)$) or to an edge that meets v (resp. $v(t)$). As $\mathcal{B}_S$ and $\mathcal{B}_{S(t)}$ are isomorphic, the difference between the decorations Λ and $\Lambda(t)$ must be in the arrows of A_1 and $A_1(t)$.

- As $I_S(f) = I_{S(t)}(f+t)$, if $I_{S\cup\{v\}}(f)$ is different to $I_{S(t)\cup\{v(t)\}}(f+t)$, the difference should be in the infinitely near points associated with the arrows attached to v and $v(t)$.

This implies that either the graphs $\mathcal{B}_{\geq v}(f)$ and $\mathcal{B}_{\geq v(t)}(f+t)$ are not isomorphic or $I_{\geq v}(f)$ is different to $I_{\geq v(t)}(f+t)$. Hence if we could prove the statement of Lemma 4.18 for any vertex v of $\mathcal{B}_f$, we would be done. We do it by induction on the floor in which v is.

If v is of the first floor then our statement is precisely Lemma 4.18. Suppose that v is of the d-th floor for $d > 1$. Let $w, v_1, ..., v_{2r-1}$ be the vertices of the first floor of $\mathcal{B}_{\geq v}(f)$ and $x_1, ..., x_{2r-1}$ their associated infinitely near points. Let ϕ be the primitive automorphism such that $I_\phi = \{x_1, ..., x_{2r-1}\}$, denote by (I_ϕ, p) its associated parental order; let $\mathcal{B}_f, (I_\phi, p), (I_\phi, p)$ the composable triple associated with f, I_f, ϕ (note that $I_\phi \subset I_f$). If we compute $\mathcal{B}_{f\circ\phi^{-1}}$ as the (I_phi, p)-composition of $\mathcal{B}_f$ and (I_ϕ, p) we see that v_{2r-1} becomes the root and v becomes a vertex of the $(d-1)$-th floor of $\mathcal{B}_{f\circ\phi^{-1}}$. By induction, for any $t \in \mathbb{C}$, there exists $v(t) \in \mathcal{B}_{f\circ\phi^{-1}+t}$ such that the graphs $\mathcal{B}_{\geq v}(f\circ\phi^{-1})$ and $\mathcal{B}_{\geq v(t)}(f\circ\phi^{-1}+t)$ are isomorphic and moreover $I_{\geq v}(f\circ\phi^{-1}) = I_{\geq v(t)}(f\circ\phi^{-1}+t)$. The vertex $v(t)$ can be viewed as a vertex of the d-th floor of $\mathcal{B}_{\geq v(t)}(f+t)$. Suppose that there exists t_0 such that either the graphs $\mathcal{B}_{\geq v}(f)$ and $\mathcal{B}_{\geq v(t_0)}(f+t_0)$ are not isomorphic, or $I_{\geq v}(f)$ is different to $I_{\geq v(t_0)}(f+t_0)$. Using the induction hypothesis we see that the differences between the graphs should be at the d-th floor, and the differences between $I_{\geq v}(f)$ and $I_{\geq v(t)}(f+t_0)$ should be at infinitely near points associated with vertices or arrows of the d-th floor. On the other hand the d-th floor of $\mathcal{B}_{\geq v(t_0)}(f+t_0)$ is the $(d-1)$-th floor of $\mathcal{B}_{\geq v(t_0)}(f\circ\phi^{-1}+t_0)$, which we have seen that is isomorphic to the $(d-1)$-th floor of $\mathcal{B}_{\geq v}(f\circ\phi^{-1})$, which is the d-th floor of $\mathcal{B}_{\geq v}(f)$; this shows that $\mathcal{B}_{\geq v}(f) \cong \mathcal{B}_{\geq v(t_0)}(f+t_0)$. The infinitely near points of $I_{\geq v}(f)$ (resp. $I_{\geq v(t_0)}(f+t_0)$) associated with the vertices and arrows of the d-th floor of $\mathcal{B}_{\geq v}(f)$ (resp. $\mathcal{B}_{\geq v(t_0)}(f+t_0)$) are determined by ϕ and the infinitely near points of $I_{\geq v}(f\circ\phi^{-1})$ (resp. $I_{\geq v(t_0)}(f\circ\phi^{-1}+t_0)$) associated with the $(d-1)$-th floor of $\mathcal{B}_{\geq v}(f\circ\phi^{-1})$ (resp. $\mathcal{B}_{\geq v(t_0)}(f\circ\phi^{-1}+t_0)$). As $I_{\geq v}(f\circ\phi^{-1}) = I_{\geq v(t_0)}(f\circ\phi^{-1}+t_0)$ we deduce that $I_{\geq v}(f)$ is equal to $I_{\geq v(t_0)}(f+t_0)$. □

REMARK 4.20. It is widely known that, due to the existence of atypical values the last Theorem is not true if we use the *embedded resolution* of the singularities at infinity, instead of using *behaviour resolutions.*

Let's finish the section addressing some directions for future research. A consequence of Theorem 4.19 is that, given a behaviour graph $\mathcal{B}$ we can view the variety $X_\mathcal{B}$ as a foliated manifold whose leaves are the orbits of the action $\rho : \mathbb{C}\times X_\mathcal{B} \to X_\mathcal{B}$ defined by $\rho(t,f) = f+t$; let $Orb(f)$ denote the orbit of f.

It is known that given any polynomial $f : \mathbb{C}^n \to \mathbb{C}$ there is a finite number of atypical values $D(f) \subset \mathbb{C}$ such that the restriction $f : \mathbb{C}^2 \setminus f^{-1}(D(f)) \to \mathbb{C} \setminus D(f)$ is a C^∞-locally trivial fibration (see [**5**]). Actually, for the case of $n = 2$ it is known that $t_0 \in D(f)$ if and only if, either $f^{-1}(t_0)$ is singular (then we will say that $t_0 \in D_{aff}(f)$), or, given t generic, the embedded resolution graphs of the singularities at infinity of $f^{-1}(t_0)$ and $f^{-1}(t)$ are different for (then we will say that $t_0 \in D_\infty(f)$). Note that the intersection $D_{aff}(f) \cap D_\infty(f)$ is not necessary empty. There have been many efforts to identify the set $D(f)$ (check [**6**] for a list of references). We can introduce two types of discriminants in $X_\mathcal{B}$:

(1) Define $\Delta_\infty(\mathcal{B}) \subset X_\mathcal{B}$ to be the set of polynomials in $f \in X_\mathcal{B}$ such that the embedded resolution graph of the singularities at infinity of $f^{-1}(0)$ is different than the embedded resolution graph of the singularities at infinity of the central fibre of a generic polynomial of $X_\mathcal{B}$.
(2) Define $\Delta(\mathcal{B})$ to be the set of polynomials $f \in X_\mathcal{B}$ that are either in $\Delta_\infty(\mathcal{B})$ or $f^{-1}(0)$ is singular .

Both discriminats are algebraic subsets of $X_\mathcal{B}$. We say that a polynomial is *generic at infinity* in $X_\mathcal{B}$ if the embedded resolution graph of the singularities at infinity of its fibre over a generic t is isomorphic to the embedded resolution graph of the singularities at infinity of the fibre over 0 of a generic polynomial of $X_\mathcal{B}$. Note that if f is generic at infinity, then $D(f) = Orb(f) \cap \Delta(\mathcal{B})$ and $D_\infty(f) = Orb(f) \cap \Delta_\infty(\mathcal{B})$. This makes very interesting the following

PROBLEM 4.21. Study the geometry of $\Delta_\infty(\mathcal{B}) \subset X_\mathcal{B}$ and $\Delta(\mathcal{B})$ and their relative position respect to the foliation in $X_\mathcal{B}$ defined by translations on the left.

If we do not want to restrict to polynomials that are generic at infinity, the we would have to define discriminants subsets inside $\Delta(\mathcal{B})$ in a similar manner.

An important invariant associated with a polynomial f, that is being increasingly studied is the monodromy representation $\pi_1(\mathbb{C} \setminus C, t_0) \to Aut(H^*(F))$, where F is its Milnor fibre (check [**16**], [**33**], [**10**]). We consider $\mathcal{C} \subset X_\mathcal{B} \times \mathbb{P}^2$ defined like in Theorem 3.28, let $\varphi : \mathcal{C} \to X_\mathcal{B}$ the restriction of the first projection; then there is a locally trivial fibration

$$\varphi : \mathcal{C} \setminus (\phi^{-1}(\Delta_\mathcal{B}) \cup X_\mathcal{B} \times L) \to X_\mathcal{B} \setminus \Delta_\mathcal{B}$$

whose restriction to $\varphi^{-1}(Orb(f))$ is $f : \mathbb{C}^2 \setminus f^{-1}(D(f)) \to \mathbb{C} \setminus D(f)$. This leads to think that the study of the monodromy representation

$$\pi_1(X_\mathcal{B} \setminus \Delta_\mathcal{B}, f - t_0) \to Aut(H^*(F))$$

associated with φ can be quite useful. For this, it is again important the study of Problem 4.21. Vassiliev has developed a method to study the topology of discriminants and complements of discriminants in various situations (see [**34**], [**35**]), it is an interesting and non trivial problem to adapt and apply this methodto our discriminants.

CHAPTER 5

The moduli problem

The rest of the memoir is devoted to the construction and study of the moduli spaces. To state precisely our moduli problem we define, for each behaviour graph $\mathcal{B}$, a functor of families. We start investigating how to (coarsely) represent these functors; this will lead us in a natural way in the realm of *algebraic groupoids*. We show how to use them to reduce the problem of the construction of moduli spaces to the study of the functors of families associated with some special representatives of the equivalence classes of behaviour graphs.

5.1. The moduli problem: groupoid approach

To define moduli spaces of polynomials with a fixed behaviour graph $\mathcal{B}$ we have to define first a moduli problem, or what is the same, define a functor $\mathcal{M}_{\mathcal{B}}$ of families of polynomials with behaviour $\mathcal{B}$. This is done as follows: let $\mathbf{Aff}_{\mathbb{C}}$ be the category of affine schemes over $\mathbb{C}$,

DEFINITION 5.1. Let n be the degree of $\mathcal{B}$, we define a functor from affine schemes to sets $\mathcal{H}_{\mathcal{B}} : \mathbf{Aff}_{\mathbb{C}} \to Sets$ by associating to each $T \in \mathbf{Aff}_{C}$ the set of polynomials $f \in \Gamma(T, \mathcal{O}_T)[X, Y]$ of the degree n, such that the morphism $T \to \mathbb{C}_n[X, Y]$ defined by the coefficients of f can be factorized through the locally closed inclusion $X_{\mathcal{B}} \hookrightarrow \mathbb{C}_n[X, Y]$.

REMARK 5.2. It is tautological that $\mathcal{H}_{\mathcal{B}}$ is represented by $X_{\mathcal{B}}$

Notice that the property of having behaviour $\mathcal{B}$ can be defined substituting the base field $\mathbb{C}$ by any algebraically closed field of characteristic 0. Then the following remark makes sense:

REMARK 5.3. If $T \in \mathbf{Aff}_{\mathbb{C}}$ is reduced then $\mathcal{H}_{\mathcal{B}}(T)$ consists of the polynomials $f \in \Gamma(T, \mathcal{O}_T)[X, Y]$ such that their pullback to any geometric point of T has behaviour graph $\mathcal{B}$.

DEFINITION 5.4. Let $\mathcal{B}$ be a behaviour at infinity of degree n, we define a moduli functor from affine schemes to sets $\mathcal{M}_{\mathcal{B}} : \mathbf{Aff}_{\mathbb{C}} \to Sets$ as follows:

$$\mathcal{M}_{\mathcal{B}}(T) = \mathcal{H}_{\mathcal{B}}(T)/\sim$$

where given $f, g \in \mathcal{H}_{\mathcal{B}}(T)$, $f \sim g$ if there exist $\phi \in Aut_T(T \times \mathbb{A}^2_{\mathbb{C}})$ such that $f \circ \phi^{-1} = g$. The functor $\mathcal{M}_{\mathcal{B}}$ associates to $T \in \mathbf{Aff}_{\mathbb{C}}$ the set of families over T of polynomials with behaviour $\mathcal{B}$ up to isomorphism.

Our aim is to construct a scheme $\mathbf{M}_{\mathcal{B}}$ such that its functor of points is as close as possible to $\mathcal{M}_{\mathcal{B}}$. Define

$$\eta : \mathbb{C}[x, y] \times Aut(\mathbb{C}^2) \to \mathbb{C}[x, y] \times \mathbb{C}[x, y] \tag{5.1}$$

by the formula $\eta(f,\phi) := (f, f\circ\phi^{-1})$. Let $\mathbb{C}[x,y]_{\leq n}$ (resp. $Aut(\mathbb{C}^2)_{\leq n}$) denote the set of polynomials (resp. automorphisms) of degree bounded by n, obviousy

$$\eta(\mathbb{C}[x,y]_{\leq n} \times Aut(\mathbb{C}^2)_{\leq m}) \subset \mathbb{C}[x,y]_{\leq n} \times \mathbb{C}[x,y]_{\leq nm}$$

For any n the set $\mathbb{C}[x,y]_{\leq n}$ is an affine space. We claim that $Aut(\mathbb{C}^2)_{\leq m}$ has a natural structure of algebraic variety: to prove the claim consider $End(\mathbb{C}^2)$ the set of algebraic endomorphisms of $\mathbb{C}^2$. The subset $End(\mathbb{C}^2)_{\leq m}$, formed by the endomorphisms of degree bounded by m, is an affine space because each element of it is given by a pair of polynomials of degree bounded by m. Let the generic element of $End(\mathbb{C}^2)_{\leq n}$ be

$$(f_1, f_2) := (\sum_{i+j\leq m} a^1_{i,j}x^iy^j, \sum_{i+j\leq m} a^2_{i,j}x^iy^j)$$

A pair (f_1, f_2) belongs to $Aut(\mathbb{C}^2)_{\leq m}$ if and only if there exists a unique pair (the inverse) of polynomials (g_1, g_2) such that $(g_1, g_2)\circ(f_1, f_2) = (x, y)$. If such an inverse exists, noting that the decompositions of (f_1, f_2) and (g_1, g_2) as chain of blowing ups and blowing downs have the same associated graph, and that this graph determines the degree, we conclude that (g_1, g_2) is of the form

$$(g_1, g_2) := (\sum_{i+j\leq m} b^1_{i,j}x^iy^j, \sum_{i+j\leq m} b^2_{i,j}x^iy^j)$$

Comparing the coefficients of each monomial of both sides of the equality

$$(g_1, g_2)\circ(f_1, f_2) = (x, y)$$

we get a linear system with variables the $b^k_{i,j}$'s and coefficients algebraic functions on the $a^k_{i,j}$'s. The pair (f_1, f_2) is in $Aut(\mathbb{C}^2)_{\leq m}$ if and only if the system has a unique solution. If we view $\{a^k_{i,j} : i + j \leq m, k \in \{1, 2\}\}$ as the coordinate functions of $End(\mathbb{C}^2)_{\leq m}$ we deduce that $Aut(\mathbb{C}^2)_{\leq m}$ is a locally closed subset on $End(\mathbb{C}^2)_{\leq m}$. This proves the claim. We can consider the composition mapping $c : End(\mathbb{C}^2) \times End(\mathbb{C}^2) \to End(\mathbb{C}^2)$ defined by $c(\phi, \psi) = \psi\circ\phi$; it is clear that the restriction $c : End(\mathbb{C}^2)_{\leq m} \times End(\mathbb{C}^2)_{\leq m} \to End(\mathbb{C}^2)_{\leq m^2}$ is a morphism, and hence also the restriction

$$c : Aut(\mathbb{C}^2)_{\leq m} \times Aut(\mathbb{C}^2)_{\leq m} \to Aut(\mathbb{C}^2)_{\leq m^2} \tag{5.2}$$

If $i : Aut(\mathbb{C}^2) \to Aut(\mathbb{C}^2)$ denotes the mapping that assigns to each automorphism its inverse. The inverse (g_1, g_2) of an automorphism (f_1, f_2) is computed by solving the linear system obtained by imposing that $(g_1, g_2)\circ(f_1, f_2) = (x, y)$. As the coefficients of the linear system are algebraic functions on the coefficients of (f_1, f_2) and the variables of the linear system are the coeficients of (g_1, g_2), using Cramer's rule, we deduce that the restriction

$$i_m : Aut(\mathbb{C}^2)_{\leq m} \to Aut(\mathbb{C}^2)_{\leq m} \tag{5.3}$$

is a morphism. Therefore the restriction

$$\eta : \mathbb{C}[x,y]_{\leq n} \times Aut(\mathbb{C}^2)_{\leq m} \to \mathbb{C}[x,y]_{\leq n} \times \mathbb{C}[x,y]_{\leq nm} \tag{5.4}$$

is a morphism.

Let $\mathcal{B}$ be a behaviour graph that is not equivalent to the behaviour graph of a polynomial defining a set of parallel lines. As all the polynomials of $X_\mathcal{B}$ have the same degree n, due to Proposition 4.17 there is a number m such that

$$\eta^{-1}(X_\mathcal{B} \times X_\mathcal{B}) \subset \mathbb{C}[x,y]_{\leq n} \times Aut(\mathbb{C}^2)_{\leq m}$$

taking into account that $X_\mathcal{B}$ is a locally closed subset of $\mathbb{C}_n[x,y]$, which in turn is a locally closed subset of $\mathbb{C}[x,y]_{\leq nm}$, and that the mapping 5.4 is a morphism we deduce that $\eta^{-1}(X_\mathcal{B} \times X_\mathcal{B})$ is locally closed in $\mathbb{C}[x,y]^3_{\leq nm}$ and that the restriction of η to $\eta^{-1}(X_\mathcal{B} \times X_\mathcal{B})$ is a morphism. We define

$$(5.5) \qquad X_{\mathcal{B},0} := X_\mathcal{B} \qquad X_{\mathcal{B},1} := \eta^{-1}(X_\mathcal{B} \times X_\mathcal{B})$$

$$(5.6) \qquad \eta_\mathcal{B} := \eta_{|X_{\mathcal{B},1}}$$

Denote by $s_\mathcal{B}$ (resp. $t_\mathcal{B}$) the composition of $\eta_\mathcal{B}$ with the first (resp. second) projection of $X_\mathcal{B} \times X_\mathcal{B}$. Given any affine scheme T and $f,g \in \mathcal{H}_\mathcal{B}(T)$ it is straightforward that $f \sim g$ if and only if there exists a morphism $u : T \to X_{\mathcal{B},1}$ such that $s_\mathcal{B} \circ u = f$ and $t_\mathcal{B} \circ u = g$. Now we are going to study $\eta_\mathcal{B} : X_{\mathcal{B},1} \to X_{\mathcal{B},0} \times X_{\mathcal{B},0}$ and give to it the structure of a so called groupoid. We will reduce the construction of moduli spaces to the problem of finding a suitable quotient by a groupoid. This has been treated quite generally by Keel and Mori in [**22**]; we will use their terminology and two of their lemmas, but not the main result of the paper because in our concrete example we can construct moduli spaces as algebraic varieties rather than algebraic spaces, as they actually do. What follows now is a brief survey of the little part of the theory of groupoids that we need for our case, following [**22**].

DEFINITION 5.5. A groupoid is a category where all the arrows are invertible, and the objects and arrows form sets.

Let $X_\bullet$ be a groupoid. Denote the set of objects by X_0 and the set of arrows by X_1. Given any arrow $\phi \in Mor(x,y)$ we say that x (resp. y) is the source (resp. target) of ϕ, this defines "source" and "target" maps $s,t : X_1 \to X_0$. Denote by $\pi : X_1 \to X_0 \times X_0$ the map having s,t as components. Notice that the composition $\phi \circ \psi$ of two arrows is defined only if $s(\phi) = t(\psi)$, this enables us to define a composition map $c : X_1 \times_{(t,s)} X_1 \to X_1$ by $c(\phi,\psi) = \psi \circ \phi$. Given any $x \in X_0$ there is a unique identity morphism in $Mor(x,x)$; this defines an "identity" map $e : X_0 \to X_1$, that turns out to be a section of both s and t. Moreover given any arrow $\phi \in Mor(x,y)$ there exists a unique inverse $\phi^{-1} \in Mor(y,x)$; define $i : X_1 \to X_1$ to be the map that takes each arrow to its inverse.

Next we define scheme-groupoids: consider the category **Sch** with the etale topology; regard any scheme X as the sheaf $Mor(-,X)$ in the etale topology; denote $Mor(T,X)$ (the so called "set of sections of X over T") by $X(T)$.

DEFINITION 5.6. A *scheme-groupoid* is the set of data given by two schemes X_0, X_1 and morphisms $s,t : X_1 \to X_0$, $c : X_1 \times_{(t,s)} X_1 \to X_1$, $e : X_0 \to X_1$, $i : X_1 \to X_1$, such that given any $T \in$ **Sch** there is a groupoid such that $X_0(T)$ and $X_1(T)$ are the sets of objects and arrows and $s(T), t(T), c(T), e(T), i(T)$ are the source, target, composition, identity and inverse maps. Presheaf-groupoids and sheaf-groupoids are defined exaxtly in the same way. (Actually a scheme-groupoid is no more than a representable presheaf-groupoid.)

EXAMPLE 5.7. (*Groupoids associated with group actions.*) Let $\mathcal{G}$ be a group scheme, Y an scheme and $\sigma : G \times Y \to Y$ an algebraic group action. Consider the morphisms $s : G \times Y \to Y$, $t : G \times Y \to Y$, $c : (G \times Y) \times_{t,s} (G \times Y) \to (G \times Y)$, $e : Y \to (G \times Y)$ and $i : (G \times Y) \to (G \times Y)$, defined by:

$$s(g,x) := x \quad t(g,x) := \sigma(g,x) \quad c((g_2, \sigma(g_1,x)),(g_1,x)) := (g_2g_1, x)$$
$$e(x) := (id_G, x) \quad i(g,x) := (g^{-1}, \sigma(g,x)).$$

Then the schemes Y and $G \times Y$, together with the morphisms s, t, c, e, i form a scheme groupoid called *groupoid associated to the group action.*

Given $\mathcal{B}$ a behaviour at infinity we can consider $X_{\mathcal{B},1}$, $X_{\mathcal{B},0}$, $s_{\mathcal{B}}$, $t_{\mathcal{B}}$. There are obviously defined identity, inverse and composition mappings $e_{\mathcal{B}} : X_{\mathcal{B},0} \to X_{\mathcal{B},1}$ $i_{\mathcal{B}} : X_{\mathcal{B},1} \to X_{\mathcal{B},1}$ $c : X_{\mathcal{B},1} \times_{t_{\mathcal{B}},s_{\mathcal{B}}} X_{\mathcal{B},1} \to X_{\mathcal{B},1}$ that are morphisms: the first for being constant when composed with the inclusion

$$X_{\mathcal{B},1} \hookrightarrow X_{\mathcal{B},0} \times Aut(\mathbb{C}^2)_{\leq m}$$

the second because the mapping 5.3 is a morphism and the third because the mapping 5.2 is a morphism.

REMARK 5.8. It is straightforward to check that $X_{\mathcal{B},0}, X_{\mathcal{B},1}, s_{\mathcal{B}}, t_{\mathcal{B}}, c_{\mathcal{B}}, e_{\mathcal{B}}, i_{\mathcal{B}}$ is a scheme-groupoid. We denote it by $X_{\mathcal{B},\bullet}$.

We give the definition of quotient that concerns us. Let $X_\bullet$ be a sheaf groupoid, then, for every scheme T, we have an equivalence relation in $X_0(T)$ defined by $\pi(T)(X(T)) \subset X_0(T) \times X_0(T)$. Consider the presheaf whose sections over T are equivalence classes in $X_0(T)$, denote by X_1/X_0 its associated sheaf in the etale topology. We say that X_1/X_0 is the quotient sheaf associated with the groupoid.

REMARK 5.9. The sheaf $X_{\mathcal{B},0}/X_{\mathcal{B},1}$ is the sheaf associated to the presheaf $\mathcal{M}_{\mathcal{B}}$ in the etale topology.

CONVENTION 5.10. By a geometric point we will understand the spectrum of an algebraically closed field that contains $\mathbb{C}$.

DEFINITION 5.11. Let $X_\bullet$ be a scheme-groupoid, Q a scheme and $q : X_0/X_1 \to Q$ a sheaf homomorphism.

(1) We say that Q is a *categorical quotient* if and only if given any scheme Q' and any sheaf homomorphism $q' : X_0/X_1 \to Q'$, there exist a unique morphism $f : Q \to Q'$ such that $f \circ q = q'$.
(2) We say that Q is a *geometric quotient* if and only if
 (i) $q(\xi) : X_0/X_1(\xi) \to Q(\xi)$ is a bijection for any geometric point ξ.
 (ii) q is submersive, i.e. if we denote also by q the composition $X_0 \to X_0/X_1 \to Q$, the $U \subset Q$ is open if and only if $q^{-1}(U)$ is open.
 (iii) The sequence of sheaves $0 \to \mathcal{O}_Q \to q_*\mathcal{O}_{X_0} \overset{s^*-t^*}{\longrightarrow} (q \circ s)_*\mathcal{O}_{X_1}$ is exact (i.e the regular functions on Q are exactly the "funtions invariant by the groupoid").

REMARK 5.12. It is known that any geometric quotient is a categorical quotient, this has been shown by Mumford in [**26**] in the context of quotients by group actions.

Let's recall the definition of *coarse moduli space* specialized to our setting:

DEFINITION 5.13. A *coarse moduli space* for the functor $\mathcal{M}_{\mathcal{B}}$ is a geometric quotient $\mathbf{M}_{\mathcal{B}}$ of the groupoid $X_{\mathcal{B},\bullet}$.

Our strategy to construct quotients involves cutting slices, for this we will need the following definition, which contains some straightforward assertions whose proof we leave to the reader.

DEFINITION 5.14. Let $X_\bullet$ a scheme-groupoid and $f : W_0 \to X_0$ a morphism. Consider the fibre product square:

$$\begin{array}{cccccc} & W_1 & \longrightarrow & X_1 & \\ (s_{W_0}, t_{W_0}) & \downarrow & & \downarrow & (s,t) \\ & W_0 \times W_0 & \stackrel{(f,f)}{\longrightarrow} & X_0 \times X_0 & \end{array}$$

define $e_{W_0}, c_{W_0}, i_{W_0}$ in the obvious way. Then we define the groupoid restricted to W_0 to be the unique groupoid having $W_0, W_1, s_{W_0}, t_{W_0}, c_{W_0}, e_{W_0}, i_{W_0}$ as objects, arrows, source, target, composition, identity and inverse.

We state a lemma that tell us in which conditions cutting slices is useful to the construction of quotients. Its proof is completely analogous to a lemma from Keel and Mori [**22**] and we do not repeat it here.

LEMMA 5.15. [**22**] *Let $X_\bullet$ a scheme-groupoid and $f : W_0 \to X_0$ a morphism, then*

(1) *The naturally defined map of sheaves $W_0/W_1 \to X_0/X_1$ is injective, and it is an isomorphism if the composition $p : W_0 \times_{(f,s)} X_1 \stackrel{proj.}{\longrightarrow} X_1 \stackrel{t}{\to} X_0$ is surjective in the etale topology.*
(2) *Assume that both s, t are universally open, that the composition p above is universally open, and as a map to its image is surjective in the etale topology. Suppose $X_0 \to Q$ is a geometric quotient. Then the image V of the composition $W_0 \to X \to Q$ is open, and the induced map $W_0 \to V$ a geometric quotient.*
(3) *Assume s, t are universally open and p surjective in the etale topology. Suppose that $q : W_0/W_1 \to Q$ defines a geometric quotient, then the map $X_0/X_1 \cong W_0/W_1 \to Q$, obtained by composing q with the inverse of the isomorphism defined in part 1 of the lemma, defines a geometric quotient.*

5.2. Comparison of groupoids

Our aim is to construct a geometric quotient of $X_{\mathcal{B},\bullet}$ for each behaviour graph $\mathcal{B}$. We will devote this section to reduce this problem to the construction of a geometric quotient of $X_{\mathcal{B},\bullet}$ for one $\mathcal{B}$ in each equivalence class of behaviour graphs.

Consider $\mathcal{B}$, $\mathcal{B}'$, equivalent behaviour graphs that are not equivalent to the behaviour of a set of parallel lines. We construct a groupoid that helps comparing the groupoids associated with $\mathcal{B}$ and $\mathcal{B}'$. Define

$$Y_0 := X_{\mathcal{B},0} \coprod X_{\mathcal{B}',0}$$

$$Y_1 := X_{\mathcal{B},1} \coprod X_{\mathcal{B}',1} \coprod \eta^{-1}(X_{\mathcal{B},0} \times X_{\mathcal{B}',0}) \coprod \eta^{-1}(X_{\mathcal{B}',0} \times X_{\mathcal{B},0})$$

$$\eta_Y := \eta_{|Y_1} : Y_1 \to Y_0 \times Y_0$$

where $\eta : \mathbb{C}[x,y] \times Aut(\mathbb{C}^2) \to \mathbb{C}[x,y] \times \mathbb{C}[x,y]$ is defined in 5.1. One shows that $\eta^{-1}(X_{\mathcal{B},0} \times X_{\mathcal{B}',0})$ and $\eta^{-1}(X_{\mathcal{B}',0} \times X_{\mathcal{B},0})$ are algebraic varieties in the same way that $X_{\mathcal{B},1}$ is shown to be algebraic for $\mathcal{B}$ not equivalent to a set of parallel lines; then η_Y is a morphism. We denote by s_Y, t_Y the composition of η_Y with each of the projections $Y_0 \times Y_0 \to Y_0$, there are obvious definitions of c_Y, e_Y and i_Y, the composition, identity element and inverse element morphisms. One checks easily that the set of data $Y_1, Y_0, s_Y, t_Y, c_Y, e_Y, i_Y$ forms an algebraic groupoid $Y_\bullet$.

DEFINITION 5.16. Given two equivalent behaviours at infinity $\mathcal{B}$ and $\mathcal{B}'$, we say that the groupoid constructed in the discussion above is the *comparison groupoid* between $\mathcal{B}$ and $\mathcal{B}'$.

We are going to need the following addendum to Theorem 3.28:

PROPOSITION 5.17. *Let $(\mathcal{B}, \leq_a)$ be an ordered behaviour graph with $n+1$ vertices. Let*

$$(5.7) \qquad Z_n(\mathcal{G}, \leq_a) \stackrel{\rho_n(\mathcal{G},\leq_a)}{\longrightarrow} \dots \stackrel{\rho_1(\mathcal{G},\leq_a)}{\longrightarrow} Z_0(\mathcal{G}, \leq_a) = X_{\mathcal{G},\leq_a} \times \mathbb{P}^2$$

be the sequence defined in Theorem 3.28. Then each $\rho_k(\mathcal{G}, \leq_a)$ is a blowing up in a smooth 2-codimensional centrum $D_k \subset Z_{i-1}(\mathcal{G}, \leq_a)$ such that, for any $(f, I_f, \alpha) \in X_{\mathcal{G},\leq_a}$, the fibre of D_k over (f, I_f, α) is the k-th infinitely near point of I_f, with reduced structure.

Let $\mathcal{E}_k$ be the exceptional divisor of $\rho_k(\mathcal{G}, \leq_a)$. If $p(v_k) = \{v_i, v_j\}$ then $D_k = \mathcal{E}_i \cap \mathcal{E}_j$ and if $p(v_k) = v_i$ then $D_k \subset \dot{\mathcal{E}}_i$. In any of the two cases, for any l such that $\max(i, j) < l < k$ there exists a section

$$s_{k,l} : X_{\mathcal{G},\leq_a} \to Z_l(\mathcal{G}, \leq_a)$$

whose image is D_k.

PROOF. There are two new statements, the first is that $\rho_k(\mathcal{G}, \leq_a)$ is a blowing up in a smooth 2-codimensional centrum. This is true because the sequence 5.7 is the pullback of the sequence of Theorem 3.14 by the smooth morphism $X_{\mathcal{B},\leq_a} \to V_{\mathcal{B},\leq_a}$. The second is the statement about the sections. It holds because of remark 3.16. □

The following lemma is a relative version of the Castelnuovo lemma for contracting $\mathbb{P}^1$'s with self-intersection -1 in an algebraic surface.

LEMMA 5.18. *Let $X \to S$ be a smooth projective morphism of relative dimension 2, where S is reduced, and $E \subset X$ a Cartier divisor. Assume that X is such that for any $s \in S$ every geometric fibre $E_s \otimes_{k(s)} \overline{k(s)}$ of the restiction $E \to S$ is isomorphic to a smooth rational curve with self-intersection -1 in $X_s \otimes_{k(s)} \overline{k(s)}$. Then there exist an S-scheme $X' \to S$, smooth and projective of relative dimension 2, and a morphism of S-schemes $p : X \to X'$ such that the restriction $p(E) \to S$ is an isomorphism and p is the blowing up of X' at $p(E)$.*

We will prove the lemma after the proof of the following theorem.

THEOREM 5.19. *If $\mathcal{B}$, $\mathcal{B}'$ are equivalent behaviours, not equivalent to the behaviour of a set of parallel lines, and there exists $M_\mathcal{B}$, a geometric quotient of $X_{\mathcal{B},\bullet}$, then there exists a geometric quotient of $X_{\mathcal{B}',\bullet}$ and it is isomorphic to $M_\mathcal{B}$.*

PROOF. Consider $Y_\bullet$, the comparison groupoid for $\mathcal{B}$ and $\mathcal{B}'$. Let i and i' be the inclusions of $X_{\mathcal{B},0}$ and $X_{\mathcal{B}',0}$ in Y_0 and let p and p' be the compositions

$$p : X_{\mathcal{B},0} \times_{i,s_Y} Y_1 \stackrel{proj.}{\longrightarrow} Y_1 \stackrel{t_Y}{\to} Y_0$$

$$p' : X_{\mathcal{B}',0} \times_{i,s_Y} Y_1 \stackrel{proj.}{\longrightarrow} Y_1 \stackrel{t_Y}{\to} Y_0$$

We start by showing that s_Y, t_Y, p and p' are universally open. We will make use of the following general notion

DEFINITION 5.20. Let $f : X \to Y$ be a morphism of schemes of finite type over $\mathbb{C}$ such that for any $x \in X$, there is $\rho : (U, u) \to (Y, f(x))$ a neigbourhood in the etale topology of $f(x)$ in Y and a morphism $s : U \to X$ such that $s(u) = x$ and $f \circ s = \rho$. Then we say that U is a morphism with *etale local sections.*

It is easy to show that any morphism with etale local sections is open and that possesing etale local sections is preserved by base change. Hence any morphism with etale local sections is universally open.

Let's show that s_Y is a morphism with etale local sections. This property is local in both source and target. We treat only the restriction of s_Y to $X_{\mathcal{B},1}$ because this case involves all the needed ideas. Note that the restriction of s_Y to $X_{\mathcal{B},1}$ is $s_{\mathcal{B}} : X_{\mathcal{B},1} \to X_{\mathcal{B},0}$. Given $(f, \phi) \in X_{\mathcal{B},1}$, let I_f the behaviour set of f. Give an admissible ordering $\leq_a$ on $\mathcal{B}$ and consider the covering morphism $\rho_{\mathcal{B},\leq_a} : X_{\mathcal{B},\leq_a} \to X_{\mathcal{B},0}$. Let (f, I, α) be a preimage by $\rho_{\mathcal{B},\leq_a}$ of f. To show the existence of an etale local section of $s_{\mathcal{B}}$ around f it is enough to show the existence of a Zariski open neighbourhood U of (f, I, α) together with a morphism $s : U \to X_{\mathcal{B},1}$ such that $s(f, I, \alpha) = (f, \phi)$ and $s_{\mathcal{B}} \circ s = \rho_{\mathcal{B},\leq_a}$.

Let $\mathcal{B}, (S, p), (S_0, p_0)$ be the composable triple associated with f, I_f, ϕ (see definition 4.6). The set S is the set of infinitely near points that are centers of the blowing up process π_ϕ given by the minimal resolution of the indeterminancy of ϕ; the set S_0 is equal to $I_f \cap S$. Consider $\mathbf{S_0} := \alpha(S_0)$, a subset of the set of vertices of $\mathcal{B}$. We will assume, to simplify the notation, that the vertices of $\mathbf{S_0}$ are the $k+1$ first vertices of $(\mathcal{B}, \leq_a)$ (the root and the first k other vertices). We consider the first k blowing ups of the sequence 5.7

$$Z_k(\mathcal{B}, \leq_a) \overset{\sigma_k(\mathcal{B},\leq_a)}{\longrightarrow} \dots \overset{\sigma_1(\mathcal{B},\leq_a)}{\longrightarrow} X_{\mathcal{B},\leq_a} \times \mathbb{P}^2. \tag{5.8}$$

If $U \subset X_{\mathcal{B},\leq_a}$ is any Zariski open subset then we denote by

$$Z_{k,U} \overset{\sigma_{k,U}}{\longrightarrow} \dots \overset{\sigma_{1,U}}{\longrightarrow} \mathbb{P}^2 \times U \tag{5.9}$$

the pullback of the sequence 5.9 to U. Let r the number of elemets in $S \setminus S_0$. We are going to construct a Zariski open neighbourhood U of (f, I, α) and a sequence of blowing ups

$$Z_{k+r,U} \overset{\sigma_{k+r,U}}{\longrightarrow} \dots \overset{\sigma_{k+1,U}}{\longrightarrow} Z_{k,U} \tag{5.10}$$

with the following properties:

(1) Each of the $\sigma_{i,U}$ of the sequence is a blowing up at a smooth centre D_i of codimension 2, such that the projection $D_i \to U$ is an isomorphism.
(2) The sequence

$$Z_{k+r,U} \overset{\sigma_{k+r,U}}{\longrightarrow} \dots \overset{\sigma_{k+1,U}}{\longrightarrow} Z_{k,U} \overset{\sigma_{k,U}}{\longrightarrow} \dots \overset{\sigma_{1,U}}{\longrightarrow} \mathbb{P}^2 \times U \tag{5.11}$$

obtained by concatenating the sequences 5.9 and 5.10, is, fibrewise over (f, I_f, α), the blowing up process π_ϕ associated with the parental system (S, p), and, fibrewise over any other $(g, I_g, \beta) \in U$, a blowing up process π whose associated parental system is isomorphic to (S, p) and, moreover, such that if we consider the automorphism ψ whose associated blowing up process is π, then the composable triples associated with (f, I_f, ϕ) and (g, I_g, ψ) are isomorphic.

We construct U and the sequence 5.10 in the following way: let $\mathcal{F}_{\mathcal{B}}$ be the tautological polynomial over $X_{\mathcal{B},\leq_a}$; let $\mathcal{C} \subset \mathbb{P}^2 \times X_{\mathcal{B},\leq_a}$ be variety defined by it (see Theorem 3.28); denote by $\tilde{\mathcal{C}}_k$ the strict transform of $\mathcal{C}$ by $\sigma_1(\mathcal{B}, \leq_a)\circ ... \sigma_k(\mathcal{B}, \leq_a)$. For any open subset $U \subset X_{\mathcal{B},\leq_a}$ denote by $\tilde{\mathcal{C}}_{k,U}$ the restriction of $\tilde{\mathcal{C}}_k$ to $Z_{k,U}$. For any $0 \leq i \leq k$ denote by $\mathcal{E}_{i,U}$ the restriction of $\mathcal{E}_i$ to $Z_{k+r,U}$, where, for $i > 1$, the divisor $\mathcal{E}_i$ is the exceptional divisor of $\sigma_i(\mathcal{B}, \leq_a)$, and $\mathcal{E}_0$ is the strict transform of $X_{\mathcal{B},\leq_a} \times L$. Let $i_1, ..., i_h$ be the integers for which $v_k \in p(v_{i_j})$; because of Proposition 5.17

$$\tilde{\mathcal{C}}_k \cap \mathcal{E}_k = \cup_{j=1}^{h} s_{i_h,k}(X_{\mathcal{B},\leq_a})$$

Let v be the biggest vertex of $S \setminus S_0$ by the parental ordering,

- If $p(v) = \{v_i, v_k\}$, the vertices v_i and v_k are connected by an edge e in $\mathcal{B}$ and there is no arrow attached to e: otherwise v would belong to S_0 because of the definition of composable triple associated with f, I_f, ϕ (definition 4.6). Therefore the sets $\tilde{C}_k$ and $\mathcal{E}_i \cap \mathcal{E}_k$ have empty intersection. Define $U := X_{\mathcal{B},\leq_a}$ and $\sigma_{k+1,U} : Z_{k+1,U} \to Z_{k,U}$ to be the blowing up of Z_k at $\mathcal{E}_i \cap \mathcal{E}_k$. Denote by $\mathcal{E}_{k+1}$ the exceptional divisor of σ_{k+1}.
- If $p(v) = \{v_k\}$ then v is an infinitely near point that have a representative that is a genuine point $x \in \dot{\mathcal{E}}_k$, the point x does not meet any of the images of the sections $s_{i_1,k}, ..., s_{i_h,k}$: otherwise v would belong to S_0 because of the definition of composable triple associated with f, I_f, ϕ. There exist a Zariski open neighbourhood U of (f, I_f, α) and a section $s_v : U \to \dot{\mathcal{E}}_{k,u}$ of the morphism $\mathcal{E}_{k,u} \to U$ satisfying first, that $s_v(f, I_f, \alpha) = x$, and second, that the image of s_v does not meet the images of the restrictions of the $s_{i_j,k}$'s to U. Define $\sigma_{k+1,U} : Z_{k+1,U} \to Z_{k,U}$ to be the blowing up of $Z_{k,U}$ at the image of the section s_v. Denote by $\mathcal{E}_{k+1}$ the exceptional divisor of σ_{k+1}.

Iterating this procedure and shrinking U when necessary we finish the construction of the sequence 5.10 so that the sequence 5.11 has the desired properties.

The morphism $Z_{k+r,U} \to U$ is smooth an projective of relative dimension 2. The total transform of $U \times L$ by the composition of the morphisms of the sequence 5.11 is a normal crossings Cartier divisor whose components are $\mathcal{E}_{i,U}$, for $i \leq r + k$. The fibre of this divisor over any point of $(g, I, \alpha) \in U$, is the total transform of L by the blowing up process given by the fibre of the sequence 5.11 over (g, I, α). For any (g, I, α) the parental system associated with this blowing up process is isomorphic to (S, p). Let $\mathcal{A}$ be the graph of (S, p). As the weight of the root of $\mathcal{A}$ is -1, the fibres of $\mathcal{E}_{0,U} \to U$ are, fibrewise over U, smooth rational curves embedded with self-intersection -1 into the fibres of $Z_{k+r,U} \to U$. Using Lemma 5.18 we see that there exists $Z'_{k+r-1,U}$, a smooth, projective U-scheme of relative dimension 2, and a morphism of U schemes $\sigma'_{k+r,U} : Z_{k+r,U} \to Z'_{k+r-1,U}$ that fibrewise over (g, I, α) is the contraction of the fibre of $\mathcal{E}_{0,U}$ over (g, I, α). The divisor $\sum_{i=1}^{k+r} \mathcal{E}_{i,U}$ is transformed by $\sigma'_{k+r,U}$ into a normal crossings Cartier divisor, that, by abuse of language, will be denoted by $\sum_{i=1}^{k+r} \mathcal{E}_{i,U}$. We can continue this procedure inductively and get a sequence

$$(5.12) \qquad Z_{k+r,U} \xrightarrow{\sigma'_{k+r,U}} \dots \xrightarrow{\sigma'_{1,U}} Z'_{0,U}$$

where each $\sigma'_{i,U}$ is the contraction of certain $\mathcal{E}_{j,U}$. The fibres of $Z'_{0,U}$ over U are each of them the result of considering first a blowing up process $\pi : X \to \mathbb{P}^2$ with graph $\mathcal{A}$, whose centers are infinitely near points to the line at infinity, and later, contracting all the possible -1 divisors of π^*L starting by L. We have seen that the result of such a procedure is isomorphic to $\mathbb{P}^2$. We conclude that the fibres of $Z'_{0,U}$ over U are all of them isomorphic to $\mathbb{P}^2$. Perhaps shrinking U, we can assume that $Z'_{0,U}$ is isomorphic to $\mathbb{P}^2 \times U$. We consider the sequence

$$(5.13) \qquad \mathbb{P}^2 \times U \overset{\sigma_{1,U}}{\longleftarrow} \dots \overset{\sigma_{k+r,U}}{\longleftarrow} Z_{k+r,U} \overset{\sigma'_{k+r,U}}{\longrightarrow} \dots \overset{\sigma'_{1,U}}{\longrightarrow} \mathbb{P}^2 \times U$$

Let $(X : Y : Z)$ be the projective reference of the $\mathbb{P}^2$ of the source $\mathbb{P}^2 \times U$. We fix a projective reference $(X' : Y' : Z')$ in the $\mathbb{P}^2$ of the target $\mathbb{P}^2 \times U$ in such a way that $Z' = 0$ defines the divisor $\mathcal{E}_{k+r,U}$ in $\mathbb{P}^2 \times U$. The projective references give us embeddings of $\mathbb{C}^2 \times U$ in both the source and target $\mathbb{P}^2 \times U$. Denote by Ψ the composition Ψ of birational maps of the last sequence. The restriction of Ψ to $\mathbb{C}^2 \times U$ is an isomorphism. Hence we can view Ψ as an automorphism of $\mathbb{C}^2 \times U$ as an U-scheme. The projective reference $(X' : Y' : Z')$ could have been chosen such that the restriction of Ψ to the fibre over (f, I_f, α) leaves fixed origin of $\mathbb{C}^2$ and is such that its differential over the origin is the identity. With this choice of projective references it is clear that if $\phi = \theta \circ H \circ \psi$ is the decomposition of ϕ in an automorphism of G, a linear automorphism and a translation then the fibre $\Psi_{(f,I_f,\alpha)}$ equals ψ. We can consider the U-automorphism of $\mathbb{C}^2$ given by $\Phi := (\theta, id_U) \circ (H, id_U) \circ \Psi$. It is clear that the fibre of Φ over U is ϕ. We have constructed Φ so that, for any $(g, I_g, \beta) \in U$, the composable triple associated with $g, I_g, \Phi_{(g,I_g,\beta)}$ is equal to the composable triple associated with (f, I_f, ϕ). As the behaviour graph of any $g \circ \Phi^{-1}_{(g,I_g,\leq_a)}$ is determined by the composable triple associated with $g, I_g, \Phi^{-1}_{(g,I_g,\beta)}$ we deduce that it is the same than the behaviour graph of $f \circ \phi$, that is $\mathcal{B}$. Define

$$s_\Phi : U \to \mathbb{C}[x, y] \times Aut(\mathbb{C}^2)$$

by the formula $s_\Phi(g, I_g, \beta) := (g, \Phi_{(g,I_g,\beta)})$. The mapping s_Φ actually defines a morphism $s_\Phi : U \to X_{\mathcal{B},1}$ that is such that $s_\Phi(f, I_f, \alpha) = \phi$ and $s_{\mathcal{B}} \circ s_\Phi = \rho_{\mathcal{B},\leq_a}$. This finishes the proof that s_Y is a morphism with local sections.

The proof that the morphisms t_Y, p and p' are also morphisms with etale local sections follows the same lines.

Now we want to show that p and p' are surjective in the etale topology. To prove that the composition

$$p : X_{\mathcal{B},0} \times_{i,s_Y} Y_1 \overset{proj.}{\longrightarrow} Y_1 \overset{t_Y}{\longrightarrow} Y_0$$

is surjective in the etale topology it is enough to show the following property: given any $f \in Y_0$, there exist an etale neighbourhood $\rho : U \to Y_0$ of f in Y_0 and a morphism $\delta : U \to X_{\mathcal{B},0} \times_{i,s_Y} Y_1$ such that $p \circ \delta = \rho$. Let $f \in Y_0$. If $f \in X_{\mathcal{B},0}$ then take ρ to be equal to $id_{X_{\mathcal{B},0}} : X_{\mathcal{B},0} \to X_{\mathcal{B},0}$ and δ be defined by the formula $\delta(g) := (g, (g, id_{\mathbb{C}^2}))$. If $f \in X_{\mathcal{B}',0}$, then, as $\mathcal{B}$ is equivalent to $\mathcal{B}'$, there exist $\phi \in Aut(\mathbb{C}^2)$ such that $f \circ \phi^{-1}$ has behaviour $\mathcal{B}$. Using the same technique than the proof that s_Y has etale local sections, one can deduce that there exists an etale neighbourhood $\rho : U \to X_{\mathcal{B}',0}$ and a morphism $\xi : U \to \eta^{-1}(X_{\mathcal{B}',0} \times X_{\mathcal{B},0})$ such that $s_Y \circ \xi = \rho$. Define $\delta : U \to X_{\mathcal{B},0} \times_{i,s_Y} Y_1$ by the formula $\delta(u) := (t_Y(\xi(u)), i_Y(\xi(u)))$, where i_Y is the inverse morphism of the groupoid $Y_\bullet$. For any $u \in U$

$$p \circ \delta(u) = t_Y(i_Y(\xi(u))) = s_Y(\xi(u)) = \rho(u)$$

and hence p is surjective in the etale topology. We can prove analogously that p' is surjective in the etale topology.

Now we can use Lemma 5.15: we let $f : W_0 \to X_0$ be the inclusion $i : X_{\mathcal{B},0} \to Y_0$. By the first part of the lemma we deduce that the sheaves $X_{\mathcal{B},0}/X_{\mathcal{B},1}$ and Y_0/Y_1 are isomorphic. Ley $\varphi : X_{\mathcal{B},0}/X_{\mathcal{B},1} \to Y_0/Y_1$ be the isomorphism. By hypothesis we know that $M_{\mathcal{B}}$ is a geometric quotient for $X_{\mathcal{B},0}/X_{\mathcal{B},1}$. Let $q : X_{\mathcal{B},0}/X_{\mathcal{B},1} \to M_{\mathcal{B}}$ the quotienting map. By the third part of the lemma we deduce that $q \circ \varphi^{-1} : Y_0/Y_1 \to M_{\mathcal{B}}$ is a geometric quotient of the groupoid $Y_\bullet$. Now take $f : W_0 \to X_0$ to be the inclusion $i' : X_{\mathcal{B}',0} \to Y_0$. By the first part of the lemma we deduce that the sheaves $X_{\mathcal{B},0}/X_{\mathcal{B},1}$ and Y_0/Y_1 are isomorphic. Let $\varphi' : X_{\mathcal{B}',0}/X_{\mathcal{B}',1} \to Y_0/Y_1$ be the isomorphism. Let $\xi : Y_0 \to M_{\mathcal{B}}$ be the composition of $Y_0 \to Y_0/Y_1 \to M_{\mathcal{B}}$. By Yoneda lemma ξ is a morphism of schemes. As the composition $\xi \circ i' : X_{\mathcal{B}',0} \to M_{\mathcal{B}}$ is surjective, by the second part of the lemma we deduce that the induced mapping $q' : X_{\mathcal{B}',0}/X_{\mathcal{B}',1} \to M_{\mathcal{B}}$ is a geometric quotient for $X_{\mathcal{B}',\bullet}$. This finishes the proof of the theorem. □

For the proof of Lemma 5.18 we will need the following auxiliary lemmas:

LEMMA 5.21. *Let $f : X \to S$ a projective morphism with S reduced and such that for any $s \in S$ the geometric fibre $X_{\bar{s}} := X_s \otimes_{k(s)} \overline{k(s)}$ is irreducible and reduced of dimension n, with n not depending on s. Then f is flat.*

PROOF. Theorem 11.8.1 from E.G.A. IV (see [**17**]) allows to reduce to the case of $S = Spec(A)$, with A discrete valuation ring. Then the result follows from Proposition III.9.7 of [**18**]. □

COROLLARY 5.22. *Let $f : X \to S$ satisfying the hypothesis of the preceding lemma. Then, f is a smooth morphism if and only if each geometric fibre is regular.*

PROOF OF LEMMA 5.18. This proof is an adaptation of the proof given in Hartshorne of the clasical Castelnovo contractibility criterion, see [**18**], Theorem V.5.7.

As $p : X \to S$ is projective we consider an S-inmersion $X \hookrightarrow \mathbb{P}^n_S$ for certain n. This induces an invertible sheaf $\mathcal{O}_X(1)$. Using Serre's vanishing theorem and composing the embedding $X \hookrightarrow \mathbb{P}^n_S$ with a d-uple embedding for certain d we can assume that $R^i p_* \mathcal{O}_X(1) = 0$ for any $i > 0$. The divisor E is embedded in $\mathbb{P}^n_S$ and, because of Lemma 5.21, it is flat over S. Hence the degree of each one of its fibres over S does not depend of the fibre. We let this degree be d. Composing with a 2-uple embedding we can assume that $d \geq 2$. We will use the invertible sheaf $\mathcal{M} := \mathcal{O}_X(1) \otimes \mathcal{L}(dE) = \mathcal{L}(H + dE)$, where H is a hyperplane of $\mathbb{P}^n_S$ to give a morphism from X to $\mathbb{P}(\mathcal{E})$ for certain locally free sheaf $\mathcal{E}$ over S.

We claim that $R^1 p_* \mathcal{L}(H + iE) = 0$ for any $0 \leq i \leq d$. If $i = 0$ the claim is true because $\mathcal{L}(H) = \mathcal{O}_X(1)$. Assume that the claim is true for $i - 1$. Consider the exact sequence of sheaves

$$0 \to \mathcal{L}(H + (i-1)E) \to \mathcal{L}(H + iE) \to \mathcal{L}(H + iE) \otimes \mathcal{O}_E \to 0$$

The associated long cohomology sequence gives

$$R^1 p_* \mathcal{L}(H + (i-1)E) \to R^1 p_* \mathcal{L}(H + iE) \to R^1 p_* (\mathcal{L}(H + iE) \otimes \mathcal{O}_E) \tag{5.14}$$

the first term of the sequence vanishes by induction hypothesis. The third term has support in E and hence if p also denotes the projection from E to S we have that$R^1 p_* (\mathcal{L}(H + iE) \otimes \mathcal{O}_E) = R^1 p_* \mathcal{L}(H + iE)_{|E}$. By hypothesis for any $s \in S$

we have that $E_{\overline{s}} \simeq \mathbb{P}^1_{\overline{k(s)}}$, and that it is embedded with self-intersection -1 in $X_{\overline{k(s)}}$. Under this isomorphism it is clear that $\mathcal{L}(H+iE)_{\overline{s}} \simeq \mathcal{O}_{\mathbb{P}^1_{\overline{k(s)}}}(d-i)$; hence, if $d \geq i$, then $H^1(E_{\overline{s}}, \mathcal{L}(H+iE)_{\overline{s}}) = 0$. Using that S is reduced we conclude that $R^1p_*(\mathcal{L}(H+iE) \otimes \mathcal{O}_E = 0$; therefore the middle term of the sequence 5.14 also vanishes. This proves the claim. A similar argument shows that $R^2p_*\mathcal{L}(H+iE) = 0$. As the relative dimension of X over S is 2, $R^jp_*\mathcal{L}(H+iE) = 0$ for $j > 2$.

As $R^ip_*\mathcal{M} = 0$ if $i > 1$ and $\mathcal{M}$ is flat over S, we conclude that $h^0(X_{\overline{s}}, \mathcal{M}_{\overline{s}})$ does not depend on s. As S is reduced this implies that $p_*\mathcal{M}$ is a locally free sheaf over S. Let $\mathcal{E} := p_*\mathcal{M}$. By Proposition II.7.12 of [**18**] if we show that the natural homomorphism $p^*\mathcal{E} = p^*p_*\mathcal{M} \to \mathcal{M}$ is surjective the we have defined an S-morphism $X \to \mathbb{P}(\mathcal{E})$. Given any $s \in S$ consider the cartesian square

$$\begin{array}{ccc} X_{\overline{s}} & \xrightarrow{i} & X \\ p_{\overline{s}} \downarrow & & \downarrow p \\ Spec(\overline{k(s)}) & \xrightarrow{j} & S \end{array} \tag{5.15}$$

To show the surjectivity it is enough to show that, for any $s \in S$, the homomorphism $i^*p^*p_*\mathcal{M} \to i^*\mathcal{M}$ is surjective. Notice that $i^*p^*p_*\mathcal{M} = p^*_{\overline{s}}j^*p_*\mathcal{M}$. As $p_*\mathcal{M}$ is locally free, using III.12.11 of [**18**], we deduce that $j^*p_*\mathcal{M}$ is naturally isomorphic to $p_{\overline{s}*}i^*\mathcal{M}$. Hence we are reduced to show that the morphism $p^*_{\overline{s}}p_{\overline{s}*}i^*\mathcal{M} \to i^*\mathcal{M}$ is surjective. This surjectivity holds because the ideal $i_*\mathcal{M}$ is known to be generated by local sections because of the proof of Theorem V.5.7. of [**18**].

We have shown that there is an S-morphism $\phi : X \to \mathbb{P}(\mathcal{E})$. Denote by X' the image of ϕ. Denote by $p' : X' \to S$ the projection morphism. Given any $s \in S$ the morphism $\phi_{\overline{s}} : X_{\overline{s}} \to X'_{\overline{s}}$ is the contraction of $E_{\overline{s}}$ defined by the classical Castelnuovo contractibility criterion (see the proof of Theorem V.5.7. of [**18**]). Hence for any $s \in S$ we know that $X'_{\overline{s}}$ is an smooth surface and we conclude from Corollary 5.22 that $X' \to S$ is smooth of relative dimension 2. The restriction $p'_{|\phi(E)} : \phi(E) \to S$ is clearly $1-1$. We would like to know that it is an isomorphism. For this we notice that given the restriction $p_{|E} : E \to S$ it is easy to construct analytic local sections around each point of S. If we compose this analytic local sections with ϕ we show that the inverse map to $p'_{|\phi(E)}$ is analytic; as we are dealing with reduced projective schemes we may conclude that it is algebraic and hence that $p'_{|\phi(E)}$ is an isomorphism.

We would finish the proof of or lemma if we show that $\phi : X \to X'$ is the blowing up of X' at $\phi(E)$. Consider $\psi : Bl_{\phi(E)}X' \to X'$. We claim that *there exits a morphism* $\varphi : X \to Bl_{\phi(E)}X'$ *such that* $\psi \circ \varphi = \phi$. Let's assume the claim. For any $s \in S$, the restriction $\varphi_{\overline{s}} : X_{\overline{s}} \to Bl_{\phi(E)_{\overline{s}}}X'_{\overline{s}}$ is an isomorphism, because of Theorem V.5.7. of [**18**]. This implies that ψ is bijective. As we know that it is proper we deduce that ψ is finite. We can deduce that ψ is an isomorphism (and hence finish the proof of the lemma modulo the claim) showing the following assertion: *Let* $f : X \to Y$ *be a finite* S*-morphism with* S *reduced such that for any point* $s \in S$ *the morphism* f_s *is an isomorphism. Then* f *is an isomorphism.* The assertion is local in S and Y. Hence we can assume that S and Y are affine. As f is finite then X is also affine. Let $X = Spec(A)$, $Y = Spec(B)$ and $S = Spec(C)$. The ring A is a finite B-module. We would finish if we show that the B-module A/B is zero. Let $\mathbf{p} \in Spec(B)$, let $\mathbf{q} = C \cap \mathbf{p}$. We have that $(A/B)_{\mathbf{p}} = A_{\mathbf{p}}/B_{\mathbf{p}}$. As $f_{\mathbf{q}}$ is an isomorphism we have $A/\mathbf{q}A = B/\mathbf{q}B$ and hence $A = B + \mathbf{q}A \subset B + \mathbf{p}A$. Localizing in $\mathbf{p}$ we obtain $A_{\mathbf{p}} \subset B_{\mathbf{p}} + (\mathbf{p}A)_{\mathbf{p}}$. By Nakayama we deduce that $(A/B)_{\mathbf{p}} = 0$. As this can be done for any $\mathbf{p} \in Spec(B)$ we deduce that $A/B = 0$.

The only thing that remains to be proved is the claim. To prove it we use the universal property of blowing up at a sheaf of ideals (see II.7.14 of [**18**]). Let $\mathcal{K}$ be the sheaf of ideals of $\phi(E)$. We need to show that $\phi^{-1}\mathcal{K} \,.\, \mathcal{O}_X$ is invertible. To save notation define $\mathcal{J} := \phi^{-1}\mathcal{K} \,.\, \mathcal{O}_X$. It is enough to show that $\mathcal{J} = \mathcal{I}$, where $\mathcal{I}$ is the ideal sheaf of E. As $\phi^{-1}(\phi(E)) \subset E$ it is clear that $\mathcal{J} \subset \mathcal{I}$. For each $s \in S$ we have that $\mathcal{J}_{\overline{s}} = \mathcal{I}_{\overline{s}}$, because $X_{\overline{s}} \to X'_{\overline{s}}$ is the blowing up of $X'_{\overline{s}}$ in $\mathcal{K}_{\overline{s}}$. From here we deduce easily that $\mathcal{J}_s = \mathcal{I}_s$. To show that $\mathcal{I}$ is equal to $\mathcal{J}$ we can work locally over X. We consider $U = Spec(A)$, an affine open subset of X such that its image in S is contained in an affine open subset $V = Spec(B)$. Let I and J be the ideals of A given by the sections of $\mathcal{I}$ and $\mathcal{J}$ over U. Let s be an arbitrary closed point of V and let $\mathbf{m}$ be its ideal in B. As $\mathcal{J}_s = \mathcal{I}_s$ we deduce that $I = J + \mathbf{m}A \cap I$. If we could show that $\mathbf{m}A \cap I = \mathbf{m}A \,.\, I$ then we would have that $I = J + \mathbf{m}A \,.\, I$. If $\mathbf{m}' \subset A$ denotes any maximal ideal containing $\mathbf{m}A$ we would have that $I = J + \mathbf{m}' . I$. Using Nakayama lemma we would conclude that $I_{\mathbf{m}'} = J_{\mathbf{m}'}$. As any maximal ideal of A lies over a maximal ideal of B and A is a reduced finitely generated $\mathbb{C}$-algebra we would deduce that $I = J$. Hence all we have to show is that $\mathbf{m}A \cap I = \mathbf{m}A \,.\, I$. Consider the exact sequence $0 \to I \to A \to A/I \to 0$ As $X \to S$ and $E \to S$ are flat we deduce that A and A/I are A-flat. This implies that I is also A-flat. Thus, we get the following diagram of exact sequences:

$$
\begin{array}{ccccccccc}
 & & 0 & & 0 & & 0 & & \\
 & & \downarrow & & \downarrow & & \downarrow & & \\
0 & \to & I \otimes_B \mathbf{m} & \to & A \otimes_B \mathbf{m} & \to & A/I \otimes_B \mathbf{m} & \to & 0 \\
 & & \downarrow & & \downarrow & & \downarrow & & \\
0 & \to & I & \to & A & \to & A/I & \to & 0 \\
 & & \downarrow & & \downarrow & & \downarrow & & \\
0 & \to & I \otimes_B B/\mathbf{m} & \to & A/\mathbf{m}A & \to & A/(I + \mathbf{m}A) & \to & 0 \\
 & & \downarrow & & \downarrow & & \downarrow & & \\
 & & 0 & & 0 & & 0 & &
\end{array}
\tag{5.16}
$$

The module of the upper right corner can be seen to be at the same time $\mathbf{m}A/I.\mathbf{m}A$ and $(I + \mathbf{m}A)/I \cong \mathbf{m}A/(I \cap \mathbf{m}A)$. This implies that $\mathbf{m}A \cap I = \mathbf{m}A \,.\, I$. □

CHAPTER 6

The moduli spaces

Given any behaviour graph $\mathcal{B}$, we will coarsely represent its associated moduli functor by constructing a geometric quotient (the moduli space) of the groupoid $X_{\mathcal{B},\bullet}$. Due to Theorem 5.19 it is enough to restrict to the case of behaviour graphs of minimal degree in their $\sim$-equivalence class. The construction of the geometric quotients is by succesive slicing of $X_{\mathcal{B},0}$, until (in most of the cases) we are reduced to the construction of the quotient of an affine variety by one or two succesive finite group actions. The finite group that appears is related in a beautiful way with the combinatorics of the graph $\mathcal{B}$. We will devote the first section on the combinatorial preliminaries that will lead to these groups.

At the end of the chapter we will describe an algorithm to decide whether two polynomials are R-equivalent, and also study the geometric properties of the moduli spaces.

6.1. Combinatorial preliminaries

During this section $\mathcal{B}$ is going to denote a behaviour graph of minimal degree.

DEFINITION 6.1. Let $\mathbf{CT}(\mathcal{B})$ be the set of composable triples $\mathcal{B},(S,p),(S_0,p_0)$ such that the (S_0,p_0)-composition of $\mathcal{B}$ and (S,p) is isomorphic to $\mathcal{B}$.

Note that if $f\in X_{\mathcal{B},0}$, each vertex of the first generation corresponds to one point at infinity of f and viceversa.

DEFINITION 6.2. If $\mathcal{B},(S,p),(S_0,p_0)$ is a composable triple then there exist a unique vertex of the first generation v such that for any $v'\in S_0$ with $v'\neq w$ we have $v'\leq_p v$. We say then that the composable triple is *emanating from v*.

LEMMA 6.3. *All non-trivial composable triples of* $\mathbf{CT}(\mathcal{B})$ *emanate from the same vertex. Moreover if* $\mathcal{B},(S,p),(S_0,p_0)$ *belongs to* $\mathbf{CT}(\mathcal{B})$*, then the graph of* (S,p) *equals* $\mathcal{A}_n$ *for certain* n.

PROOF. Suppose that there exist two composable triples $\mathcal{B},(S,p),(S_0,p_0)$ and $\mathcal{B},(S',p'),(S_0',p_0')$ in $\mathbf{CT}(\mathcal{B})$ emanating from different vertices v and v'. Consider $f\in X_{\mathcal{B},0}$. Because of Lemma 4.7 and Proposition 4.10 there is an automorphism of ϕ (resp. ψ) in G such that $\mathcal{B},(S,p),(S_0,p_0)$ (resp. $\mathcal{B},(S',p'),(S_0',p_0')$) is the composable triple associated with f,I_f,ϕ (resp. f,I_f,ψ) and $f\circ\phi^{-1}$ (resp $f\circ\psi^{-1}$) have behaviour graph $\mathcal{B}$. Let p and p' be the points at infinity of f that corresponds to the vertices of the first generation v and v'. As $v\neq v'$ we have that $p\neq p'$. If we consider the extension of ϕ (resp. ψ) to a birational transformation of $\mathbb{P}^2$ then p (resp. p') is the indetermination point of ϕ and of ϕ^{-1} (resp. ψ and ψ^{-1}). As $p\neq p'$ if $\phi\in Aut(\mathbb{C}^2)_{n_1,\ldots,n_r}$ and $\psi\in Aut(\mathbb{C}^2)_{m_1,\ldots,m_s}$ we have that $\psi\circ\phi^{-1}\in Aut(\mathbb{C}^2)_{n_r,\ldots,n_1,m_1,\ldots,m_s}$. As we have the sequence of degrees $\deg(f\circ\phi^{-1})=\deg(f)=\deg(f\circ\psi^{-1})$ we are contradicting Corollary 4.15.

For the second assertion we consider an automorphism ϕ in G such that the composable triple associated with f, I_f, ϕ is $\mathcal{B}, (S,p), (S_0, p_0)$. If the graph of (S,p) were $\mathcal{A}_{n_1,\dots,n_r}$ for $r > 1$ we also would contradict Corollary 4.15. □

Our aim now is to study the set $\mathbf{CT}(\mathcal{B})$ for each behaviour graph of minimal degree.

6.1.1. Graphs with more than one vertex of the first generation. In this subsection $\mathcal{B}$ will denote a behaviour graph of minimal degree with more than one vertex of the first generation. Let I denote the set of vertices of $\mathcal{B}$.

Because of Remark 4.5, if $\mathcal{B}, (S,p), (S_0, p_0)$ is in $\mathbf{CT}(\mathcal{B})$ then $(S,p), (S_0, p_0) \subset I$ and the (S_0, p_0)-composition of $\mathcal{B}$ and (S,p) is the graph that results of letting the last vertex of S become the root of $\mathcal{B}$. This makes very easy to look for the elements of $\mathbf{CT}(\mathcal{B})$: look for the vertices of $\mathcal{B}$ such that if we move the root to them we get a graph isomorphic to $\mathcal{B}$.

DEFINITION 6.4. We define $Aut_{rm}(\mathcal{B})$, the group of *root moving automorphisms* of $\mathcal{B}$ to be the group of automorphisms of the weighted decorated non rooted graph obtained from $\mathcal{B}$ by letting the root be a non distinguished vertex.

LEMMA 6.5. *Given an admissible ordering $\leq_a$ on $\mathcal{B}$ there is a naturally defined injection*

$$\Omega : \mathbf{CT}(\mathcal{B}) \to Aut_{rm}(\mathcal{B})$$

We do not make $\leq_a$ appear in the notation of Ω because we will always refer to the same admissible ordering.

PROOF. Consider a composable triple $\mathcal{B}, (S,p), (S_0, p_0)$ belonging to $\mathbf{CT}(\mathcal{B})$. We explain how to construct its image under Ω. Let $\{v_0, ..., v_{2n-1}\}$ be the vertices of S ordered decreasingly by the parental ordering. Denote by (I, p_I) the parental system associated with the graph $\mathcal{B}$. We know that $S = S_0 \subset I$. Consider the branches of $\mathcal{B}$ (see Definition 4.2). We say that two of this branches are combinatorially equivalent if and only if there exist an automorphism of $\mathcal{B}$ taking one into the other. We will denote by $\sim_c$ the combinatorial equivalence relation in the set of branches. The admissible ordering induces a total order in each of the $\sim_c$-classes of branches that is inherited from the total order in the $\sim_c$-classes of their respective maxima.

Let $\mathcal{B}_c$ denote the (S_0, p_0)-composition of $\mathcal{B}$ and (S,p). The graph $\mathcal{B}_c$ is the graph obtained from $\mathcal{B}$ by letting v_{2n-1} become the root. By hypothesis there exists an isomorphism $\phi : \mathcal{B} \to \mathcal{B}_c$. As isomorphisms preserve roots we conclude that $\phi(v_0) = \phi(v_{2n-1})$. The vertex v_1 has the property that $p(v_1) = \{v_0\}$. If p_c denotes the parental function of $\mathcal{B}_c$ then $p_c(\phi(v_1))$ should be equal to $\{\phi(v_0)\} = \{v_{2n-1}\}$. The vertex v_1 satisfyies $p_c(v_1) = \{v_{2n-1}\}$. We claim that $\phi(v_1) = v_1$. Suppose that $\phi(v_1) \neq v_1$, then, because of the shape of the parental system (S,p) we have that $S \cap \phi(S) = \{v_{2n-1}\}$. Hence $S \cup \phi(S)$ is a decreasing linearly ordered chain in $\mathcal{B}$ of length $2(2n-1)$ starting from the root. Then $S'' := \phi(S) \cup \phi^2(S)$ is a linearly ordered chain of length $2(2n-1)$ in $\mathcal{B}_c$ starting from v_{2n-1}. As the free central vertex of the floor $\phi(S)$ is not v_1 and any vertex of $S'' \setminus \{v_{2n-1}\}$ is smaller or equal than it by the parental order we deduce that $S'' \cap S = \{v_{2n-1}\}$; then $S \cup S''$ is a linear chain of length $3(2n-1)$ in $\mathcal{B}$ starting at v_0. Repeating this we can find chains of vertices of lenght $s(2n-1)$ for arbitrary s which is a contradiction. We have thus proved that $\phi(v_1) = v_1$. The vertex v_2 is the only vertex such that $p(v_2) = \{v_0, v_1\}$;

as v_{2n-2} is the only vertex whose image by p_c is $\{v_{2n-1}, v_1\} = \{\phi(v_0), \phi(v_1)\}$ we deduce that $\phi(v_2) = v_{2n-2}$. Analogously we prove that $\phi(v_i) = v_{2n-i}$ for $2 \leq i \leq n$.

Let x denote a vertex in $\{v_0, ..., v_n\}$ or an edge connecting two of these vertices. Let $\mathcal{R}$ be a branch attached to x; if there are l branches attached to x in the isomorphism class of $\mathcal{R}$ then there are also l branches attached to $\phi(x)$ isomorphic to $\mathcal{R}$. The admissible ordering $\leq_a$ induces a total order in both sets of branches. Let $R_1^1, ..., R_l^1$ and $R_1^2, ..., R_l^2$ be the ordered sets of branches attached to x and $\phi(x)$ respectively. The admissible ordering, together with the canonical ordering of $\mathcal{B}$ induces a total ordering in $\mathcal{B}$, and by restriction in each of the R_j^i's. We can define an isomorphism from R_j^1 to R_j^2 for any $j \leq l$ by sending the i-th vertex (resp. arrow) of R_j^1 to the i-th vertex (resp. arrow) of R_j^2.

We define an element of $\psi \in Aut_{rm}(\mathcal{B})$ as follows: let $\psi(v_i) = \phi(v_i)$ for any $0 \leq i \leq n$, let $\psi(v_{n+i}) = v_{n-i}$ if $0 \leq i \leq n-2$, let $\psi(v_{2n-1}) = v_0$. This defines the restriction of ψ to S. It only remains to define it at the branches. Let $\mathcal{R}$ be a branch attached to a vertex in $\{v_0, ..., v_n\}$ or to an edge connecting two of this vertices, if $\mathcal{R} = R_i^1$ in the notations of the paragraph above we define the restriction of ψ to $\mathcal{R}$ to be equal to the isomorphism from $\mathcal{R}_i^1$ to $\mathcal{R}_i^2$. If $\mathcal{R}$ is a branch attached to a vertex in $\{v_n, ..., v_{2n-1}\}$ or to an edge connecting two of this vertices, then $\mathcal{R}$ should be equal to some R_i^2 in the notations of the paragraph above; we define the restriction of ψ to $\mathcal{R}$ to be equal to the inverse of the isomorphism from $\mathcal{R}_i^1$ to $\mathcal{R}_i^2$. This finishes the definition of ψ. We define the image of $\mathcal{B}, (S,p), (S_0, p_0)$ by Ω to be ψ.

The injectivity of Ω is because the composable triple $\mathcal{B}, (S,p), (S_0, p_0)$ is determined by the position of the last vertex of S, and this position is just the image of the root of $\mathcal{B}$ by $\Omega(\mathcal{B}, (S,p), (S_0, p_0))$. □

We extend the mapping Ω in the following way:

DEFINITION 6.6. Let $\mathcal{B}$ and $\leq_a$ as in the lemma above, define

$$\Theta : Aut(\mathcal{B}) \times \mathbf{CT}(\mathcal{B}) \to Aut_{rm}(\mathcal{B})$$

by the formula $\Theta(\alpha, \mathcal{B}, (S,p), (S_0, p_0)) := \Omega(\mathcal{B}, (S,p), (S_0, p_0)) \circ \alpha$

PROPOSITION-DEFINITION 6.7. *The mapping Θ of the last definition is a bijection. We will denote its inverse by Ξ.*

PROOF. The injectivity is easy. For the surjectivity consider an arbitrary $\psi \in Aut_{rm}(\mathcal{B})$. Denote by w the root of $\mathcal{B}$. Let $v = \psi(w)$. If $v = w$ then ψ belongs to $Aut(\mathcal{B})$ and is clearly in the image of Θ; so we can assume $v \neq w$. Let p denote the parental function in $\mathcal{B}$ and consider $S := \{v' \in \mathcal{B} : v' \geq_p v\}$. The induced parental system (S, p) is linearly ordered by the parental order. As $S \subset \mathcal{I}$ we deduce that, for certain $n_1, ..., n_r$, the parental system (S, p) is isomorphic to a parental subsystem of the parental system corresponding to the graph $\mathcal{A}_{n_1,...,n_r}$. We claim that v is the last vertex of a floor of $\mathcal{A}_{n_1,...,n_r}$. To prove the claim we use the following fact: given a parental graph $\mathcal{G}$ we say that a vertex is an *ending vertex* of $\mathcal{G}$ if it has weight -1 and is not the root; deletion of an ending vertex gives rise to another parental graph. Let (I, p) be the parental system associated with the graph $\mathcal{B}$, consider the subset $I \setminus S$ and define the branches of $\mathcal{B}$ as in Definition 4.2. Notice that we can successively delete ending vertices of the branches until we get a graph $\mathcal{A}$ whose set of vertices is S. Consider the graph $\mathcal{B}_c$ obtained from $\mathcal{B}$ by moving the root to v. As $\psi \in Aut_{rm}(\mathcal{B})$ takes the root to v we deduce that $\mathcal{B}_c$ is isomophic to $\mathcal{B}$ as rooted graphs. We can successively delete from $\mathcal{B}_c$ all the vertices corresponding to the branches of $\mathcal{B}$, because the available ending vertices at each stage do not change with the moving of the root from w to v. After these deletions we get a graph $\mathcal{A}_c$ that is the result of moving the root of $\mathcal{A}$ from w to v. As $\mathcal{B}_c$ is isomorphic to $\mathcal{B}$, any graph obtained by deletions of ending vertices of $\mathcal{B}_c$ should be the graph of a parental system contained in $\mathcal{I}$, and this is clearly not the case for $\mathcal{G}_c$ if v is not either equal to w, the case that we have already excluded, or the last vertex of a floor of $\mathcal{A}_{n_1,\dots,n_r}$. This proves the claim.

The claim shows that (S,p) is isomorphic to the parental system of the graph $\mathcal{A}_{n_1,\dots,n_s}$ for certain $s \leq r$. Consider the composable triple $\mathcal{B},(S,p),(S,p)$. As $\psi \in Aut_{rm}(\mathcal{B})$ takes the root to v and the (S,p)-composition of $\mathcal{B}$ and (S,p) is the result of taking the root of $\mathcal{B}$ to v we deduce that $\mathcal{B},(S,p),(S,p)$ is in $\mathbf{CT}(\mathcal{B})$. Let $\phi := \Theta(id,\mathcal{B},(S,p),(S,p))$. It is clear that $\phi^{-1}\circ\psi \in Aut(\mathcal{B})$. Then $\psi = \Theta(\phi^{-1}\circ\psi,\mathcal{B},(S,p),(S,p))$. This proves the surjectivity. □

6.1.2. Graphs with only one vertex of the first generation. In this subsection $\mathcal{B}$ will denote a minimal degree behaviour graph that has exactly one vertex of the first generation. Consider $\mathcal{B},(S,p),(S_0,p_0) \in \mathbf{CT}(\mathcal{B})$ and define the branches of $\mathcal{B}$ as in Definition 4.2. Consider the set of all the vertices v of S_0 such that either v is the attaching point of a branch or an arrow, or v meets an edge that is the attaching point of a branch or an arrow. Let x be the vertex that, among them, is closest to the root of $\mathcal{B}$ (i.e. the vertex such that the number of edges that we need to connect it with the root is as small as possible).

LEMMA 6.8. *There exists a positive number r with the following property: given $\mathcal{B},(S,p),(S_0,p_0)$ and x as above, if the graph of (S,p) is $\mathcal{A}_n$ for $n > r$, and the vertices of S are $w,v_1,\dots,v_{2n-1}$ then*

- *If $r = 1$ then x is the root, there is no branch or arrow attached to v_0 and there is an arrow attached to the edge between v_0 and v_2*
- *If $r > 1$ then $x = v_r$, there is no branch or arrow attached to v_i for $i \leq r$ and there is an arrow attached to the edge between v_r and v_{r+1}*

PROOF. We will use polynomials to prove this purely combinatorial statement. Let $f \in X_{\mathcal{B}}$. As $\mathcal{B}$ has only one vertex of the first generation f has only one point p at infinity. Consider the infinite blowing up process whose first blowing up center is p and whose parental system is $(\{w\}\cup\{v_i\}_{i\in\mathbb{N}},p)$, where $p(v_1) = w$, $p(v_2) = \{v_0,v_1\}$ and $p(v_i) = \{v_1,v_{i-1}\}$ for $i \geq 3$. Let E_i be the exceptional divisor corresponding to the blowing up π_i at v_i, L the strict transform of the line at infinity and $\tilde{C}_i$ the strict transform of the curve defined by f at each stage of the blowing up process. Let r be the smallest integer such that $\tilde{C}_{r+1}$ meets the total transform of L at a point different from $E_1 \cap E_{r+1}$. We claim that this point has to be $E_r \cap E_{r+1}$ if $r > 2$ or $E_2 \cap E_0$ if $r = 1$. Suppose the contrary; then $\tilde{C}$ meets $\dot{E}_{r+1}$ at a point q. Consider $\pi'_{r+2} : X_{i+2} \to X_{r+1}$ to be the blowing up at q. Choose blowing ups $\pi'_i : X_i \to X_{i-1}$ for $r+3 \leq i \leq 2(r+1)-1$ such that the blowing up process

$$X_{2(r+1)-1} \xrightarrow{\pi'_{2(r+1)-1}} \dots \xrightarrow{\pi'_{r+2}} X_{r+1} \xrightarrow{\pi_{r+1}} \dots \xrightarrow{\pi_1} \mathbb{P}^2 \tag{6.1}$$

has graph $\mathcal{A}_{r+1}$. Consider the automorphism $\psi \in G$ associated with the blowing up process above. The strict transform of C by the last blowing up process meets the exceptional divisor corresponding to a free lateral vertex, and it does not meet any exceptional divisor corresponding either to a fixed lateral vertex or to the root. This implies that $\deg(f \circ \phi^{-1}) < \deg(f)$, contradicting the fact that $\mathcal{B}$ is of minimal degree. This proves the claim, from which the lemma follows easily. □

Let (I, p) be the parental system of $\mathcal{B}$. Consider an admissible ordering $\leq_a$ on it and let $\{w, v_1, ..., v_n\}$ be the set of vertices of $\mathcal{B}$ ordered by the total ordering in $\mathcal{B}$. We know that the set of vertices of I bigger than or equal to v_r is $\{w, v_1, ..., v_r\}$ and that the parental function on them is given by $p(v_1) = \{w\}$, $p(v_2) = \{w, v_1\}$ and $p(v_i) = \{v_1, v_{i-1}\}$ for $2 \leq i \leq r$. We also know that $\{w, v_2, ..., v_{r-1}\}$ are not among the images under p of other vertices different from $v_1 ..., v_r$, and, moreover, no of them and no edge joinig two of them supports decoration. This implies that the weights of the vertices $\{w, v_2, ..., v_{r-1}\}$ are such that we can successively delete the vertices $\{w, v_2, ..., v_{r-1}\}$ and that, moreover, by doing so we do not modify the decoration.

DEFINITION 6.9. We define the *modified graph associated with* $\mathcal{B}$, and denote it by $\mathcal{B}^m$, to be the rooted weighted decorated graph obtained from $\mathcal{B}$ by deleting $\{w, v_2, ..., v_{r-1}\}$ and defining v_r to be the new root.

Given the modified graph $\mathcal{B}^m$ of a graph $\mathcal{B}$ we notice that the weight of the root of $\mathcal{B}^m$ is always -1. We can recover the graph $\mathcal{B}$ by considering a linear chain of vertices of lenght r and weights $-1, -2, ..., -2$ and identifying the last one with v_r. Given $\alpha \in Aut(\mathcal{B})$, as $w, ..., v_{r-1}$ should be left fixed, the automorphism α induces an automorphism by restriction on $\mathcal{B}^m$. Conversely , given $\alpha' \in Aut(\mathcal{B}^m)$ we define an automorphism of $\alpha \in Aut(\mathcal{B})$ by letting the restriction of α to $\mathcal{B}^m$ be α' and letting the rest of the vertices fixed.

REMARK 6.10. There is a canonical isomorphism $Aut(\mathcal{B}) \cong Aut(\mathcal{B}^m)$. Any admissible ordering of $\mathcal{B}$ induces an admissible ordering on $\mathcal{B}^m$ in a canonical way, and viceversa.

Let $\mathcal{B}, (S, p), (S_0, p_0)$ be a composable triple such that the (S_0, p_0)-composition of $\mathcal{B}$ and (S, p) is isomorphic to $\mathcal{B}$. The graph of (S, p) is $\mathcal{A}_s$ for certain s. Consider a composable triple for which $s > r$. Denote by $\mathcal{B}_c$ the (S_0, p_0)-composition of $\mathcal{B}$ and (S, p). If the vertices of (S, p) ordered by $\leq_p$ are $\{w, v_1, ..., v_{2s-1}\}$ then $S_0 = S \setminus \{v_{2s-r+1}, ..., v_{2s-1}\}$ and $S_1 = S \setminus \{w, v_2, ..., v_{r-1}\}$ (review the construction of (S_0, p_0)-compositions for the definition of S_1). Let the branches of $\mathcal{B}$ be as in Definition 4.2. Denote by $\mathcal{G}$ the result of attaching all the branches and decoration of $\mathcal{B}$ to the graph $\mathcal{A}_s$ of (S, p). Then $\mathcal{B}_c^m$ is the result of deleting from it the vertices $\{w, v_2, ..., v_{r-1}\} \cup \{v_{2s-r+1}, ..., v_{2s-1}\}$ and letting v_{2s-r} be the root. On the other hand $\mathcal{B}^m$ is, by definition, the result of deleting from $\mathcal{G}$ the same and letting the root be v_r. Consequently $\mathcal{B}_c^m$ is the result of moving the root of $\mathcal{B}^m$ to v_{2s-r}. We have proved:

LEMMA 6.11. *Let $\mathcal{B}, (S, p), (S_0, p_0)$ a composable triple such that the graph of (S, p) is $\mathcal{A}_s$ for $s > r$, and such that the (S_0, p_0)-composition of $\mathcal{B}$ and (S, p) is isomorphic to $\mathcal{B}$. Then the vertex $v_{2s-r} \in S$ actually belongs to $\mathcal{B}^m$ and $\mathcal{B}_c^m$ is the result of moving the root of $\mathcal{B}^m$ to v_{2s-r}.*

DEFINITION 6.12. We define $\mathbf{CT}(\mathcal{B}^m)$ as the set of composable triples $\mathcal{B}, (S, p)$, (S_0, p_0) such that the graph of (S, p) is $\mathcal{A}_s$ for $s > r$ and the (S_0, p_0)-composition of $\mathcal{B}$ and (S, p) is isomorphic to $\mathcal{B}$.

From this point the combinatorics that we are going to work out are quite similar to the last subsection. We will show in detail how to adapt the proof of the next lemma and be sketchy in the rest.

LEMMA 6.13. *Given a behaviour graph $\mathcal{B}$ of the type that we are considering and an admissible ordering $\leq_a$ on it there is a naturally defined injection*

$$\Omega : \mathbf{CT}(\mathcal{B}^m) \to Aut_{rm}(\mathcal{B}^m)$$

PROOF. Given a composable triple $\mathcal{B}, (S, p), (S_0, p_0)$ belonging to $\mathbf{CT}(\mathcal{B})$ we explain how to construct its image under Ω. We use the terminology of the discussion preceding Lemma 6.11. We say that two branches of $\mathcal{B}$ are $\sim_c$ equivalent if and only if there exist an automorphism of $\mathcal{B}$ taking one into the other. The admissible ordering induces a total order in each of the isomorphism classes of branches that is inherited from the total order in the $\sim_c$-classes of their respective maxima.

The graph $\mathcal{B}_c$ is the graph obtained from $\mathcal{G}$ by letting v_{2s-1} become the root and deleting $\{w, v_2, ..., v_{r-1}\}$. By hypothesis there exists an isomorphism $\varphi : \mathcal{B} \to \mathcal{B}_c$. As isomorphisms preserve roots we conclude that $\varphi(w) = \varphi(v_{2s-1})$. The vertex v_1 satisfies that $p(v_1) = \{w\}$. If p_c denotes the parental ordering of $\mathcal{B}_c$ then $p_c(\varphi(v_1))$ should be equal to $\{\varphi(w)\} = \{v_{2n-1}\}$ and hence it is of the first generation. As we are dealing with graphs with only one vertex of the first generation then $\varphi(v_1) = v_1$. This implies that $\varphi(v_i) = v_{2s-i}$ for $2 \leq i \leq n$.

The rest of the proof is parallel to the last three paragraphs of the proof of Lemma 6.5. Only the proof of the injectivity is slightly different: the injectivity is because the composable triple $\mathcal{B}, (S, p), (S_0, p_0)$ is determined by the position of the vertex of v_{2s-r}, and this position is just the image of the root of $\mathcal{B}^m$ by $\Omega(\mathcal{B}, (S, p), (S_0, p_0))$. □

We extend the mapping Ω in the following way:

DEFINITION 6.14. Let $\mathcal{B}$ and $\leq_a$ as in the lemma above, define

$$\Theta : Aut(\mathcal{B}^m) \times \mathbf{CT}(\mathcal{B}^m) \to Aut_{rm}(\mathcal{B}^m)$$

by the formula $\Theta(\alpha, \mathcal{B}, (S, p), (S_0, p_0)) := \Omega(\mathcal{B}, (S, p), (S_0, p_0)) \circ \alpha$.

PROPOSITION-DEFINITION 6.15. *The mapping Θ of the last definition is a bijection. We will denote its inverse by Ξ.*

PROOF. It is an easy adaptation of the proof of Proposition-Definition 6.7. □

6.2. Construction of the moduli spaces

Cutting appropiate slices we will construct geometric quotients for the groupoid associated to any graph $\mathcal{B}$ of minimal degree. If one only aims to contruct the geometric quotients, then one can choose a simpler way of cutting the slices that the one that will be presented. We have choosen a more complicated way of cutting the slices in order to be able to prove Theorem 6.49, which explains the geometry of the slices. We are going to consider different cases which we list here:

- Case A: the set $\mathbf{CT}(\mathcal{B})$ has an element different from the trivial one.
- Case B: the set $\mathbf{CT}(\mathcal{B})$ has no element different from the trivial one, and $\mathcal{B}$ is not equivalent to the behaviour of a set of parallel lines.
- Case C: the graph $\mathcal{B}$ is equivalent to the behaviour of a set of parallel lines.

6.2.1. A preliminary lemma. Let $\pi : X \to \ldots \to \mathbb{P}^2$ be a blowing up process with graph $\mathcal{A}_{n,m}$; denote by $\pi_i : X_i \to X_{i-1}$ the i-th blowing up of π (then $X_0 = \mathbb{P}^2$); by x_i and E_i the blowing up center and the exceptional divisor of π_i; let $E_0 := L$, the line at infinity of $\mathbb{P}^2$; denote by p_i the vertex of $\mathcal{A}_{n,m}$ associated with E_i. We will assume that $x_1 := (0 : 1 : 0)$.

Let $(X : Y : Z)$ be the projective reference that fixes the embedding $\mathbb{C}^2 \subset \mathbb{P}^2$ by imposing $(\mathbb{C}^2, (x, y)) = (\mathbb{P}^2 \setminus L, (X/Z, Y/Z))$.

Consider the affine chart $(U_Y, (x, z))$ with $U_Y := \mathbb{P}^2 \setminus \{Y = 0\}$ and $(x, z) := (X/Y, Z/Y)$. As its origin is $(0 : 1 : 0)$ the restriction $\pi_1 : \pi_1^{-1}(U_Y) \to U_Y$ is the blowing up of this affine chart at the origin, and $\pi^{-1}(U_Y)$ is covered the two standard blowing up charts. The center x_2 is the origin of the first standard chart, which will be denoted $(U_1, (u_1, v_1))$. The restriction $\pi_2 : \pi_2^{-1}(U_1) \to U_1$ is the blowing up at the origin of U_1, hence $\pi_2^{-1}(U_1)$ can be covered by the two standard blowing up charts, and, if $n > 2$, the center x_3 of π_3 is at the origin of the second of them, which will be called $(U_2, (u_2, v_2))$. Continuing this procedure inductively we can locate x_i at the origin of the second standard blowing up chart $(U_{i-1}, (u_{i-1}, v_{i-1}))$ of π_{i-1} as long as $3 \leq i \leq n$, and define $(U_i, (u_i, v_i))$ to be the second blowing up chart of π_i.

The point x_{n+1} belongs to the two standard blowing up charts of π_n and it is not the origin of any of them. Consider $(U_n, (u_n, v_n))$, the second blowing up chart of π_n; then the exceptional divisor E_n is defined by $v_n = 0$ and x_{n+1} will have coordinates $(d_{n+1}, 0)$. We consider new coordinate fuctions $u'_n := u_n - d_{n+1}$ and $v'_n := v_n$, which define a new affine chart $(U_n, (u'_n, v'_n))$ of X_n. Now the restriction $\pi_{n+1} : \pi_{n+1}^{-1}(U_n) \to U_n$ is the blowing up at the origin of the affine chart $(U_n, (u'_n, v'_n))$ of X_n. Iterating this procedure, for any $n < i \leq 2n$, we can locate x_i in the second standard blowing up chart $(U_{i-1}, (u_{i-1}, v_{i-1}))$ of the blowing up π_{i-1} at the origin x_{i-1} of the chart $(U_{i-2}, (u'_{i-2}, v'_{i-2}))$; then x_i will have coordinates $(d_i, 0)$ respect to (u_{i-1}, v_{i-1}); we will define the chart $(U_{i-1}, (u'_{i-1}, v'_{i-1}))$ by $u'_{i-1} := u_{i-1} - d_{i-1}$ and $v'_{i-1} = v_{i-1}$.

Up to now we have defined charts $(U_i, (u_i, v_i))$ of X_i for any i such that the vertex p_i is in the first floor of $\mathcal{A}_{n,m}$. We define the charts $(U_i, (u_i, v_i))$ of X_i when p_i is in the second floor of $\mathcal{A}_{n,m}$ analogously than when p_i is in the firt floor, but letting $(U_{2n-1}, (u'_{2n-1}, v'_{2n-1}))$ play the role of $(U_Y, (x, z))$.

NOTATION 6.16. Let E_i be such that the vertex v_i is not free central. The *canonical chart* (V_i, x) of E_i is defined by $V_i := E_i \cap U_i$ and $x := u_{i|V_i}$.

LEMMA 6.17. *Let $\phi \in Aut(\mathbb{C}^2)$. With the notations of the last discussion, for any $i \leq 2(n+m-1)$ the mapping ϕ lifts to a birational transformation $\phi_i : X_i \dashrightarrow X_i$, and moreover*

- *If $\phi(x, y) = (ax, by)$, then the expression of $\phi_{n|E_i}$ respect to the canonical chart (V_n, x) is $\phi_n(x) = a^n b^{1-n} x$.*
- *If $\phi(x, y) = (x + \lambda, y)$, then $\phi_{i|E_i} = id_{|E_i}$ if v_i is fixed and of the first floor, and the the expression of $\phi_{n+1|E_{n+1}}$ respect to the canonical chart (V_{n+1}, x) is $\phi_{n+1}(x) = x + A\lambda$ where A is a non zero constant depending algebraically on $\pi \in V_{\mathcal{A}_{n,m}}$.*
- *If $\phi(x, y) = (x, y + \lambda x^r)$ with $1 \leq r < n$, then $\phi_{i|E_i} = id_{|E_i}$ if $i < 2n - r$, and the expression of $\phi_{2n-r|E_{2n-r}}$ respect to the canonical chart (V_{2n-r}, x)*

is $\phi_{2n-r}(x) = x + B_r\lambda$ *where* B_r *is a non zero constant depending algebraically on* $\pi \in V_{\mathcal{A}_{n,m}}$.

- *If* $\phi(x,y) = (x, y+\lambda)$, *then* $\phi_{i|E_i} = id_{|E_i}$ *if* $i < 2n+m$ *and* v_i *is not free central, and the the expression of* $\phi_{2n+m|E_{2n+m}}$ *respect to the canonical chart* (V_{2n+m}, x) *is* $\phi_{2n+m}(x) = x + B_0\lambda$ *where* B_0 *is a non zero constant depending algebraically on* $\pi \in V_{\mathcal{A}_{n,m}}$.

PROOF. The proof is an straightforard but tedious computation. □

6.2.2. Case A. The set $\mathbf{CT}(\mathcal{B})$ has an element different from the trivial one.

Let $\leq_a$ be an admissible ordering on $\mathcal{B}$ and let $\leq_t$ be the total order associated with it. Denote by $\leq_p$ the parental order of $\mathcal{B}$. We will assume for simplicity that the only vertex from which the composable triples of $\mathbf{CT}(\mathcal{B})$ can emanate is the first (biggest) vertex of the first generation. Consider the morphism $\rho_{\mathcal{B},\leq_a} : X_{\mathcal{B},\leq_a} \to X_{\mathcal{B},0}$ and let $W^0_\bullet$ denote the pullback of the groupoid $X_{\mathcal{B},\bullet}$ by this morphism. Using Lemma 5.15 we can reduce the construction of a geometric quotient of $X_{\mathcal{B},\bullet}$ to the construction of a geometric quotient of $W^0_\bullet$. Because of Definition 5.14, an arrow of $W^0_\bullet$ is a pair

$$(((f, I_f, \alpha), (g, I_g, \beta)), \phi) \in (W^0_0 \times W^0_0) \times_{(X_{\mathcal{B},0}\times X_{\mathcal{B},0})} X_{\mathcal{B},1} \tag{6.2}$$

such that $g = f\circ\phi^{-1}$. Now we are going to cut successively smaller slices of $W^0_\bullet$ until we force ϕ to be of a special form. Let us fix some notations that will be useful: assume that we have already cut i slices and that, by succesive pullback we have got a groupoid $W^i_\bullet$. The $i+1$-th slice will be a Zariski closed subset $j^{i+1} : W^{i+1}_0 \hookrightarrow W^i_0$. We will define the groupoid $W^{i+1}_\bullet$ to be the pullback groupoid of $W^i_\bullet$ by the last inclusion (see Definition 5.14). The source and target maps of $W^{i+1}_\bullet$ are called s^{i+1} and t^{i+1} and we will denote by p^{i+1} the composition

$$p^{i+1} : W^{i+1}_0 \times_{j^{i+1},s^i} W^i_1 \xrightarrow{proj} W^i_1 \xrightarrow{t^i} W^i_0$$

STEP 1: There exist a morphism $\sigma : W^0_0 \to L$ that takes $(f, I_f, \alpha) \in W^0_0$ to the point at infinity of f that corresponds to the first vertex of the first generation. Define $W^1_0 := \sigma^{-1}(0:1:0)$.

LEMMA 6.18. *The variety* W^1_0 *is affine.*

PROOF. Define a morphism $\phi : V_{\mathcal{B},\leq_a} \to L$, that forgets from any set of infinitely near points all except the first one of them. The morphism σ is the composition of $\tau_{\mathcal{G},\leq_a}$ with ϕ. The fact that the mophism $\tau_{\mathcal{G},\leq_a}$ is affine is a simple corollary of the geometric study that has been made about it. The morphism ϕ is the composition of the morphisms of the sequence 3.12; each of these morphisms is affine, and hence ϕ also. We deduce that σ is an affine morphism. This proves the lemma. □

By Lemma 5.15, if we show that s^0, t^0 and p^1 are universally open and that p^1 is surjective in the etale topology, then we have reduced the construction of a geometric quotient of $W^0_\bullet$ to the construction of a geometric quotient of $W^1_\bullet$. We know already that s^0 and t^0 are universally open (they have Zariski local sections). We prove that p^1 is surjective in the etale topology: consider $(f, I_f, \alpha) \in W^0_0$, there exists $H \in GL(2,\mathbb{C})$ such that $H(\sigma(f, I_f, \alpha)) = (0:1:0)$. Hence if $\beta : \mathcal{B}_{f\circ H^{-1}} \to \mathcal{B}$ is an isomorphism then $(f\circ H^{-1}, I_{f\circ H^{-1}}, \beta) \in W^1_0$. Notice that (f, I_f, α) is the image under p^1 of the element:

$$((f\circ H^{-1}, I_{f\circ H^{-1}}, \beta), (((f, I_{f\circ H^{-1}}, \beta), (f, I_f, \alpha)), H^{-1})) \in W^1_0 \times_{j^1,s^0} W^0_1$$

This proves the surjectivity at the level of points. To show that p^1 is surjective in the etale topology we extend the construction above to a Zariski open neighbourhood of (f, I_f, α): if $p \in L$ is different than $\sigma(f, I_f, \alpha)$, then over the Zariski open subset $\sigma^{-1}(L \setminus \{p\})$ there exists a family of linear automorphisms of $\mathbb{P}^2$

$$\mathcal{H} : \sigma^{-1}(L \setminus \{p\}) \times \mathbb{P}^2 \to \sigma^{-1}(L \setminus \{p\}) \times \mathbb{P}^2$$

such that for any $(g, I_g, \alpha) \in \sigma^{-1}(L \setminus \{p\})$ we have that

$$\mathcal{H}_{(g,I_g,\alpha)}(\sigma(g, I_g, \alpha)) = (0 : 1 : 0)$$

Using this it is easy to show the existence of a section of p^1 over $\sigma^{-1}(L \setminus \{p\})$. This shows the surjectivity of p^1 in the Zariski topology, and hence in the etale topology.

Now we show that s^1 is universally open (it will be needed for further cutting of slices), for this we will show that it is a morphism with Zariski local sections. Let

$$a := \big(((f, I_f, \alpha), (g, I_g, \beta)), \phi\big) \in W_1^1$$

As s^0 is a morphism with Zariski local sections, there exists a neighbourhood U of (f, I_f, α) in W_0^0 and a section $\xi : U \to W_1^0$ such that $\xi(f, I_f, \alpha) = a$. The surjectivity of p^1 in the Zariski topology implies that there exist a Zariski open neighbourhood V of (g, I_g, β) and a section $\eta : V \to W_1^0$ of s^0 such that

$$\eta(g, I_g, \beta) = \big(((g, I_g, \beta), (g, I_g, \beta)), id\big)$$

and $t^0 \circ \eta(V) \subset W_0^1$. The Zariski local section of s^1 that we are looking for is the morphism $\mu : (t^0 \circ \xi)^{-1}(V) \cap U \cap W_0^1 \to W_1^1$ defined by the formula

$$\mu(h, I_h, \gamma) := \eta(t_0(\xi(h, I_h, \gamma)) \centerdot \xi(h, I_h, \gamma)$$

where "$\centerdot$" denotes the multiplication of composable arrows in the groupoid $W_\bullet^0$. The proof that t^1 and p^1 are universally open is similar.

Let $a := \big(((f, I_f, \alpha), (g, I_g, \beta)), \phi\big) \in W_1^1$. Decompose ϕ as $\theta \circ H \circ \psi$, where $\psi \in G$, H is linear and θ is a translation . As $\mathcal{B}$ is of minimal degree, because of Corollary 4.15 we deduce that ψ must be primitive. As $(f, I_f, \alpha) \in W_0^1$ the indetermination point of ψ must be $(0 : 1 : 0)$ and hence ψ must be triangular. As $(g, I_g, \beta)) \in W_0^1$ we deduce that $H(0 : 1 : 0) = (0 : 1 : 0)$. We have shown the following: there is a decomposition $\phi = \theta \circ H \circ \psi$ where

$$\begin{aligned} \psi(x, y) &= (x, y + \textstyle\sum_{k=2}^{r} a_k x^k) \\ H(x, y) &= (ax, by + cx) \\ \theta(x, y) &= (x + d, y + e) \end{aligned} \tag{6.3}$$

From this moment we need to distinguish three subcases:

Case A.1. Assume that there exist at least three vertices of the first generation in $\mathcal{B}$.

STEP 2. We introduce a new slice $W_0^2 \subset W_0^1$ by defining W_0^2 to be the closed subset of W_0^1 formed by the triples (f, I_f, α) such that the points at infinity of f corresponding to the second and third vertices of the first generation are $(1 : 0 : 0)$ and $(1 : 1 : 0)$. As W_0^2 is an closed subset of the affine W_0^1, then it is also affine. As in the last step one proves that we can reduce the construction of a geometric quotient of $W_\bullet^1$ to the construction of a geometric quotient of the induced groupoid $W_\bullet^2$.

Let $a := \big(((f, I_f, \alpha), (g, I_g, \beta)), \phi\big) \in W_1^2$. Decompose ϕ as $\theta \circ H \circ \psi$, where ψ, H and θ are like in formulae 6.3. As $\theta \circ H$ is an affine automorphism it induces a canonical bijection between $I_{f\circ\psi^{-1}}$ and $I_{f\circ\psi^{-1}\circ(\theta\circ H)^{-1}}$; this bijection translates the total ordering of the points at infinity of $f\circ\psi^{-1}$ induced by β into a total ordering of the points at infinity of $f\circ\psi^{-1}\circ(\theta\circ H)^{-1}$. As the first point at infinity of $f\circ\psi^{-1}$ is $(0:1:0)$ there is a unique automorphism H_1 of the form $H_1(x,y)=(x,by+cx)$ such that the images of the first, second and third points at infinity of $f\circ\psi$ are $(0:1:0)$, $(1:0:0)$ and $(1:1:0)$ respectively. We decompose H as $H=H_2\circ H_1$. As H_1 leaves the first, second and third points at infinity in its desired place and θ leaves fixed the line at infinity we conclude that H_2 also should do it, hence H_2 should be of the form

$$H_2(x,y) \quad = \quad (\lambda x, \lambda y) \tag{6.4}$$

STEP 3. Let $\{w, v_1, ..., v_n\}$ be the set of vertices of $\mathcal{B}$ ordered decreasingly respect to the total order associated with $\leq_a$. Let v_k be the biggest vertex of $(\mathcal{B}, \leq_a)$ such that there exist $(\mathcal{B},(S,p),(S_0,p_0)) \in \mathbf{CT}(\mathcal{B})$ for which v_k is a free lateral vertex of (S_0,p_0). Then the vertex $v_{k'} := p(v_k)$ should be the fixed central vertex of (S_0,p_0). Consider the set $\{w, v_{i_1}, ..., v_{i_s} = v_{k'}\}$ of vertices of $(\mathcal{B}, \leq_a)$ that are bigger or equal than $v_{k'}$ by $\leq_p$. The restriction of the parental function to it is as follows: $p(v_{i_j}) = \{v_{i_1}, v_{i_{j-1}}\}$ for $3 \leq j \leq s$, $p(v_{i_2}) = \{w, v_{i_1}\}$, $p(v_{i_1}) = \{w\}$. Any $(f, I_f, \alpha) \in W_0^2$, together with the sequence $\{w, v_{i_1}, ..., v_{i_s}\}$ determines a saturated set of infinitely near points $\{x_{i_1}, ..., x_{i_s}\} \subset I_f$, let $X_s \xrightarrow{\pi_{i_s}} ... \xrightarrow{\pi_{i_1}} \mathbb{P}^2$ be the associated blowing up process. Let E_{i_s} be the exceptional divisor associated with π_{i_s}. By Lemma 6.17 the restriction of the lifting of H_2 to $E_{i_s} \subset X_s$ respect to the canonical chart of E_{i_s} is of the form

$$H_2(x) = \lambda x \tag{6.5}$$

As the infinitely near point $x_k \in I_f$ associated with v_k belongs to $\dot{E}_{i_s}$ we have that $x(x_k) \neq 0$. Define W_0^3 to be the closed subset of W_0^2 formed by the (f, I_f, α)'s for which $x(x_k) = 1$. It is affine for being a closed subset of an affine variety.

Equation 6.5 implies that given any $(f, I_f, \alpha) \in W_0^2$, composing f with transformations of the form $H_2(x,y) = (\lambda x, \lambda y)$, we can move the point x_k within the whole $\dot{E}_{i_s}$. From here imitating the argument of step 1 it is not difficult that the quotienting can be reduced to $W_\bullet^3$.

Let $a := \big(((f, I_f, \alpha), (g, I_g, \beta)), \phi\big) \in W_1^3$. Decompose ϕ as $\theta\circ H_2\circ H_1\circ\psi$, where ψ, H_1, H_2 are like in step 2 and θ is like in formulae 6.3. What are the restrictions that the cutting of the last slice imposes to H_2 and θ?.

By Lemma 6.17, the coordinate $x(x_k)$ of x_k respect to the canonical chart of E_s is not affected when we compose f with a translation. As we have $x(x_k) = 1$, by equation 6.5 we conclude that there is a unique possible choice of H_2. Now it is easy:

LEMMA 6.19. *If $a := \big(((f, I_f, \alpha), (g, I_g, \beta)), \phi\big)$ is an arrow of W_1^3 and ϕ is decomposed as $\theta\circ H_2\circ H_1\circ\psi$ as above, then H_1 and H_2 are determined by (f, I_f, α) and ψ. Moreover H_1 and H_2 depend algebraically on (f, I_f, α) and ψ.*

STEP 4. The shape of $\mathcal{B}$ implies that if $(\mathcal{B},(S,p),(S_0,p_0)) \in \mathbf{CT}(\mathcal{B})$ then $S = S_0$. Hence there exist a vertex v_l such that $p(v_l) = v_k$. We can assume v_l to be the biggest with this property. Given $(f, I_f, \alpha) \in W_0^3$ there is an infinitely near point $x_l \in I_f$ associated with v_l. Let $\{w, v_{i_1}, ..., v_{i_s}, v_{i_{s+1}} = v_k\}$ be the vertices of

$\mathcal{B}$ bigger or equal than v_k by $\leq_p$. Let

$$X_s \xrightarrow{\pi_{i_s}} \dots \xrightarrow{\pi_{i_1}} \mathbb{P}^2$$

be the blowing up process considered in last step and $\pi_{i_{s+1}} : X_{s+1} \to X_s$ be the blowing up of X_s at x_k; denote by $E_{i_{s+1}}$ the exceptional divisor of $\pi_{i_{s+1}}$. Then $x_l \in E_k$. We define the closed subset $W_0^4 \subset W_0^3$ as the set of $(f, I_f, \alpha) \in W_0^3$ such that the coordinate of x_l respect to the canonical chart (V_{s+1}, x) of $E_{i_{s+1}}$ is $x(x_l) = 0$.

Suppose that we have a translation of the form $\theta_1(x, y) = (x + d, y)$. By Lemma 6.17 the restriction of the lifting of θ to E_k respect of the canonical chart is of the form

$$\theta(x) = x + Ad \tag{6.6}$$

where A is a non zero constant that depends algebraically of the location of the centers of the blowing up process that gives rise to E_k, and, hence, that depends algebraically of I_f. We deduce that given any $(f, I_f, \alpha) \in W_0^3$, composing with a translation of the form above we can move the point x_l to a point whose coordinate respect to the canonical chart is 0. Now it is easy to see that the quotienting of $W_\bullet^3$ can be reduced to the quotienting of $W_\bullet^4$.

Let $a := \big(((f, I_f, \alpha), (g, I_g, \beta)), \phi\big) \in W_1^4$. Decompose ϕ as $\theta \circ H_2 \circ H_1 \circ \psi$, where ψ, H_1, H_2 are like in step 2 and θ of he form $\theta(x, y) = (x + d.y + e)$, like in formulae 6.3. We have proved that H_1 and H_2 are determined by (f, I_f, α) and ψ. On the other hand fixing the coordinate $x(x_l)$ respect to the canonical chart of E_k implies that the parameter d is also determined. By Lemma 6.17 the coordiante $x(x_l)$ does not change when we compose f with any translation of the form $\theta(x, y) = (x, y + e)$. Hence there are no restrictions over the parameter e.

Step 5. We have to consider three types of graphs:

Type 1: There exists a free vertex $v_m \in \mathcal{B}$ such that $p(v_m)$ is free, and such that the only vertex v_d of the first generation such that $v_d \geq v_m$ is not v_1, the first vertex of the first generation of $\mathcal{B}$ respect to $\leq_t$. We choose v_m to be the greatest among the vertices with its property respect to the total order. We will assume for simplicity that $v_d = v_2$, the second vertex of the first generation, and hence, that its associated point of I_f for any $(f, I_f, \alpha) \in W_0^4$ is $(1 : 0 : 0)$. Let $w, v_2 = v_{j_1}, \dots, v_{j_r} = v_m$ be the vertices greater of equal than v_m by $\leq_p$. Then the parental system restricted to $w, \dots, v_{j_{r-1}}$ is isomorphic to the restriction of parental system of the graph $\mathcal{A}_{r-2}$ to the vertices greater or equal than the first free lateral vertex.

Type 2: The graph $\mathcal{B}$ is not of type 1, and there exists v_m, a free vertex such that $p(v_m)$ is free of the second floor. We can consider v_m maximal with its property in $(\mathcal{B}, \leq_a)$. Let $w, v_{j_1}, \dots, v_{j_r} = v_m$ the vertices greater ot equal than $p(v_m)$ by $\leq_p$. Then there are integers n, m such that the parental system restricted to $w, v_{j_1}, \dots, v_{j_{r-1}}$ is isomorphic to the restriction of parental system of the graph $\mathcal{A}_{n,m}$ to the vertices greater or equal than the first free lateral vertex of second floor. As $\mathcal{B}$ is not of type 1 we deduce that $v_{j_1} = v_1$, the first vertex of the first generation.

We define the slice for types 1 and 2: given any $(f, I_f, \alpha) \in W_0^4$ let x_m and x_{j_i} be the points of I_f associated to v_m and v_{j_i} for $i \leq r$ (notice that $x_{j_1} = (1 : 0 : 0)$ for type 1 and $x_{j_1} = (0 : 1 : 0)$ for type 2). Consider $Y_r \xrightarrow{\pi_{j_r}} \dots \xrightarrow{\pi_{i_1}} \mathbb{P}^2$ the blowing up process whose centers are the x_{j_i}'s; let $E_{j_{r-1}}$ be the exceptional divisor of $\pi_{j_{r-1}}$.

Then $x_m \in \dot{E}_{j_{r-1}}$. We define the closed subset $W_0^5 \subset W_0^4$ as the set of (f, I_f, α)'s such that the coordinate of x_m respect to the canonical chart (V_k, x) of $E_{j_{r-1}}$ is $x(x_m) = 0$. Suppose that we have a translation of the form $\theta_1(x, y) = (x, y + e)$. Lemma 6.17 (interchanging the role of the coordinates (x, y) for type 1) implies that the restriction of the lifting of θ_1 to E_{j_r} respect of the canonical chart is of the form

$$\theta_1(x) = x + Ae \tag{6.7}$$

where A is a non zero constant depending algebraically on I_f . Hence given any $(f, I_f, \alpha) \in W_0^4$, composing with a translation of the form above we can move the point x_m to a point such that $x(x_m) = 0$. As before one can see that the quotienting of $W_\bullet^4$ can be reduced to the quotienting of $W_\bullet^5$.

Let $a := \big(((f, I_f, \alpha), (g, I_g, \beta)), \phi\big) \in W_1^4$. Decompose ϕ as $\theta \circ H_2 \circ H_1 \circ \psi$, where ψ, H_1, H_2 and θ are like in the last step. We have seen that H_1, H_2 and d are determined by (f, I_f, α) and ψ. Now, as we fix the coordinate $x(x_m)$ respect to the canonical chart of E_{i_r}, Lemma 6.17 implies that e is also determined.

Type 3: The graph $\mathcal{B}$ is neither of Type 1 nor of Type 2. In this case Lemma 6.17 imply that, given any $(f, I_f, \alpha) \in W_0^4$ and any translation of the form $\theta(x, y) = (x, y + e)$, we have that $I_{f \circ \theta^{-1}} = I_f$.

Denote the point $(0 : 1 : 0)$ by p. If $(f, I_f, \leq_a) \in W_0^4$ then the intersection multiplicity $k = I_p(C, L)$ is determined by $\mathcal{B}$. Then $a_{k,n-k} \neq 0$. Define W_0^5 as closed subset of W_0^4 formed by the (f, I_f, α)'s such that if $f(x, y) = \sum_{i+j \leq n} a_{i,j} x^i y^j$ then $a_{k,n-k-1} = 0$.

If $(f, I_f, \leq_a) \in W_0^4$ then $a_{k,n-k-1}/k$ is the unique value of e such that the composition $f \circ \theta^{-1}$ has $a_{k,n-k-1} = 0$, for θ of the form $\theta(x, y) = (x, y + e)$. Now is easy to show that we can reduce the quotienting of $W_\bullet^4$ to the quotienting of $W_\bullet^5$.

Let $a := \big(((f, I_f, \alpha), (g, I_g, \beta)), \phi\big) \in W_1^4$. Decompose ϕ as $\theta \circ H_2 \circ H_1 \circ \psi$, where ψ, H_1, H_2 are like in step 2 and θ of the form $(\theta(x, y) = (x + d, y + e)$. In last step we saw that H_1, H_2 and the parameter d are determined by (f, I_f, α) and ψ. Now we have seen that the parameter e also.

We finally define the groupoid that we are going to be able to quotient for any $\mathcal{B}$ of case B.1, whatever the type is:

$$W_{\mathcal{B},\bullet} := W_\bullet^5 \tag{6.8}$$

In the previous discussion we have shown

LEMMA 6.20. *If $a := \big(((f, I_f, \alpha), (g, I_g, \beta)), \phi\big)$ is an arrow of $W_{\mathcal{B},1}$ and ϕ is decomposed as $\theta \circ H_2 \circ H_1 \circ \psi$ as above, then H_1, H_2 and θ are determined by (f, I_f, α) and ψ. Moreover H_1, H_2 and θ depend algebraically on (f, I_f, α) and ψ.*

DEFINITION 6.21. Define $Z_{\mathcal{B}} \subset V_{\mathcal{B},\leq_a}$ to be the closed subset of $V_{\mathcal{B},\leq_a}$ formed by sets of infinitely near points such that

(1) the points corresponding to the first three vertices of the first generation are $(0 : 1 : 0)$, $(1 : 0 : 0)$ and $(1 : 1 : 0)$ respectively.
(2) If x_k, x_l are the infinitely near points corresponding to the vertices v_k and v_l considered in steps 3 and 4, then their coordinates respect to the corresponding canonical charts are $x(x_k) = 1$ and $x(x_l) = 0$.

(3) If x_m is the infinitely near point corresponding to the vertex v_m considered in types 1 and 2 of step 5, then its coordinate respect to the corresponding canonical chart is $x(x_m) = 1$.

REMARK 6.22. If $\mathcal{B}$ is of type 1 or 2 then $W_0^5 = \tau_{\mathcal{B},\leq a}^{-1}(Z_{\mathcal{B}})$. If $\mathcal{B}$ is of type 3 then $W_0^4 = \tau_{\mathcal{B},\leq a}^{-1}(Z_{\mathcal{B}})$ and, moreover, if θ is of the form $\theta(x,y) = (x, y+e)$ and $(f, I_f, \alpha) \in W_0^4$ then $I_{f\circ\theta^{-1}} = I_f$.

Case A.2. The graph $\mathcal{B}$ has exactly two vertices of the first generation.

STEP 2. We introduce the slice $W_0^2 \subset W_0^1$ by defining W_0^2 to be the closed subset of W_0^1 formed by the triples (f, I_f, α) such that the point at infinity of f corresponding to the second vertex of the first generation is $(1:0:0)$. The fact that W_0^2 is affine and that we can reduce the construction of a geometric quotient of $W_\bullet^1$ to the construction of a geometric quotient of the induced groupoid $W_\bullet^2$ is shown like in subcase A.1.

Let $a := \big(((f, I_f, \alpha), (g, I_g, \beta)), \phi\big) \in W_1^2$. Decompose ϕ as $\theta\circ H\circ\psi$, with ψ, H and θ like in formulae 6.3. There is a unique automorphism H_1 of the form $H_1(x,y) = (x, y+cx)$ such that the images of the first and second points at infinity of $f\circ\psi$ are $(0:1:0)$ and $(1:0:0)$ respectively. We decompose H as $H = H_2\circ H_1$. As H_1 leaves the first and second points at infinity in its desired place and θ leaves fixed the line at infinity we conclude that H_2 leaves $(0:1:0)$ and $(1:0:0)$ fixed; hence it is of the form

$$H_2(x,y) = (\lambda x, \mu y) \tag{6.9}$$

STEP 3. The third slice W_0^3 is defined in the same way than in subcase A.1, and also as in that case it is proved that the quotienting problem can be reduced to the construction of a geometric quotient of $W_\bullet^3$.

Consider $a = \big(((f, I_f, \alpha), (g, I_g, \beta)), \phi\big) \in W_1^3$. Decompose ϕ as $\theta\circ H_2\circ H_1\circ\psi$ like in the last step. Let α be the coordinate respect to the canonical chart of E_{i_s} of the point of $I_{f\circ(H_1\circ\psi)^{-1}}$ corresponding to v_k. Define $H_{2,1}(x,y) = (\alpha^{-1}x, \alpha^{-1}y)$, then the coordinate of the point of $I_{f\circ(H_{2,1}\circ H_1\circ\psi)^{-1}}$ corresponding to v_k respect to the canonical chart of E_{i_s} is 1. Decompose H_2 as $H_2 = H_{2,2}\circ H_{2,1}$ where $H_{2,1}(x,y) = (\alpha^{-1}x, \alpha^{-1}y)$ and $H_{2,2}(x,y) = (\lambda x, \mu y)$ for certain λ, μ. The restriction to E_{i_s} of the lifting of $H_{2,2}$ to X_{i_s} respect to canonical chart is $H_{2,2}(x) = (\lambda^s/\mu^{s-1})x$: hence $a \in W_1^3$ if and only if

$$\lambda^s = \mu^{s-1} \tag{6.10}$$

We define the set

$$\mathbf{H}_{\mathcal{B}}^1 := \{(\lambda, \mu) : \lambda^s = \mu^{s-1}\} \tag{6.11}$$

The set $\mathbf{H}_{\mathcal{B}}^1$, with pairwise multiplication, is a group isomorphic to $\mu_s \times \mathbb{C}^*$, where μ_s denotes the group of s-th roots of unity. The isomorphism is $\gamma : \mu_s \times \mathbb{C}^* \to \mathbf{H}_{\mathcal{B}}^1$, defined by $\gamma(\xi, t) := (\xi t^{s-1}, t^s)$. We have shown that $a \in W_0^3$ if and only if $(\lambda, \mu) \in \mathbf{H}_{\mathcal{B}}^1$.

STEPS 4 AND 5. In this steps we proceed exactly like in step 4 and 5 of subcase A.1.

Like in subcase A.1 we define

$$W_{\mathcal{B},\bullet} := W_\bullet^5 \tag{6.12}$$

The following lemma follows from the slicing procedure

LEMMA 6.23. *If* $a := \big(((f, I_f, \alpha), (g, I_g, \beta)), \phi\big)$ *is an arrow of* $W_{\mathcal{B},1}$ *and* ϕ *is decomposed as* $\theta \circ H_{2,2} \circ H_{2,1} \circ H_1 \circ \psi$ *as above, then:*

(1) *The automorphisms* H_1 *and* $H_{2,1}$ *are determined by* (f, I_f, α), ψ *and depends algebraically on them.*
(2) *The automorphism* $H_{2,2}$ *is of the form* $H_{2,2}(x, y) = (\lambda x, \mu y)$ *with* $(\lambda, \mu) \in \mathbf{H}^1_{\mathcal{B}}$.
(3) *The automorphism* θ *is determined by* (f, I_f, α), ψ, $H_{2,2}$ *and depends algebraically on them.*

DEFINITION 6.24. Define $Z_{\mathcal{B}} \subset V_{\mathcal{B},\leq_a}$ to be the closed subset of $V_{\mathcal{B},\leq_a}$ formed by sets of infinitely near points such that satisfy properties 2 and 3 of Definition 6.21 and, in addition, the following condition:

(1) the points corresponding to the first two vertices of the first generation are $(0 : 1 : 0)$ and $(1 : 0 : 0)$ respectively.

REMARK 6.25. If $\mathcal{B}$ is of type 1 or 2 then $W_0^5 = \tau^{-1}_{\mathcal{B},\leq_a}(Z_{\mathcal{B}})$. If $\mathcal{B}$ is of type 3 then $W_0^4 = \tau^{-1}_{\mathcal{B},\leq_a}(Z_{\mathcal{B}})$ and, moreover, if θ is of the form $\theta(x, y) = (x, y + e)$ and $(f, I_f, \alpha) \in W_0^4$ then $I_{f \circ \theta^{-1}} = I_f$.

Now, for some time, we will treat simultaneously cases A.1 and A.2 in order to construct the quotient of $W_{\mathcal{B},\bullet}$, at the end of the construction some differences will arise and we will treat each case separately again.

We are going to prove that the groupoid $W_{\mathcal{B},\bullet}$ is actually a group action. To be able to treat cases A.1 and A.2 simultaneously, we define $\mathbf{H}^1_{\mathcal{B}}$, if $\mathcal{B}$ is of case A.1, to be the trivial group. Now for both cases we can define the group $\mathbf{G}_{\mathcal{B}}$ to be

$$\mathbf{G}_{\mathcal{B}} := Aut_{rm}(\mathcal{B}) \times \mathbf{H}^1_{\mathcal{B}} \tag{6.13}$$

where the multiplication in $Aut_{rm}(\mathcal{B})$ is $\beta_1 \bullet \beta_2 := \beta_1 \circ \beta_2$. We define a mapping

$$\rho : \mathbf{G}_{\mathcal{B}} \times W_{\mathcal{B},0} \to W_{\mathcal{B},1} \tag{6.14}$$

in the following way: consider $(f, I_f, \alpha) \in W_{\mathcal{B},0}$ and $(\beta, (\lambda, \mu)) \in \mathbf{G}_{\mathcal{B}}$. Let $\Xi(\beta) = (\gamma, \mathcal{B}, (S, p), (S_0, p_0))$. Note that, as we are dealing with a graph $\mathcal{B}$ with more than one vertex of the first generation, the parental systems (S, p) and (S_0, p_0) must be equal. Define $(S_f, p_f) := (\alpha^{-1}(S, p))$. Consider the composable triple $\mathcal{B}_f, (S_f, p_f), (S_f, p_f)$. Following the reasoning of the proof of Lemma 4.7 we see that there exists a unique automorphism $\psi \in G$ such that the composable triple associated with (f, I_f, ψ) is $\mathcal{B}_f, (S_f, p_f), (S_f, p_f)$ (the existence is the statement of the lemma and the unicity is because $S_f \subset I_f$). The arguments of the proof of Theorem 5.19 show that ψ depends algebraically on $(f, I_f, \alpha), \beta$.

The graph $\mathcal{B}_{f \circ \psi^{-1}}$ can be identified with the result of letting the last vertex of S_f become the root of $\mathcal{B}_f$, this last vertex is $\alpha^{-1}(\beta(w))$, where w is the root of $\mathcal{B}$. Under this identification the isomorphism $\beta^{-1} \circ \alpha : \mathcal{B}_{f \circ \psi^{-1}} \to \mathcal{B}$ is an isomorphism of rooted graphs. Now we consider two cases:

- The graph $\mathcal{B}$ is of type $A.1$: in the construction of $W_{\mathcal{B},\bullet}$ for this case we have shown that there exists a unique affine isomorphism A such that

$$(f \circ (A \circ \psi)^{-1}, I_{f \circ (A \circ \psi)^{-1}}, \beta^{-1} \circ \alpha) \in W_{\mathcal{B},0}$$

  (Notice that the graph $\mathcal{B}_{f \circ (A \circ \psi)^{-1}}$ can be canonically identified with $\mathcal{B}_{f \circ \psi}$ for being A affine).

- The graph $\mathcal{B}$ is of type $A.2$: in the construction of $W_{\mathcal{B},\bullet}$ for this case we have shown that there exists unique H_1 and $H_{2,1}$ of the form $H_1(x,y) = (x, y+bx)$ and $H_{2,1}(x,y) = (\gamma x, \gamma y)$ such that

$$(f\circ(H_{2,1}\circ H_1\circ\psi)^{-1}, I_{f\circ(H_{2,1}\circ H_1\circ\psi)^{-1}}, \beta^{-1}\circ\alpha) \in W_0^3$$

Moreover we have shown that if $H_{2,2}(x,y) = (\lambda x, \mu y)$ then there exists a unique translation θ such that

$$(f\circ(\theta\circ H_{2,2}\circ H_{2,1}\circ H_1\circ\psi)^{-1}, I_{f\circ(\theta\circ H_{2,2}\circ H_{2,1}\circ H_1\circ\psi)^{-1}}, \beta^{-1}\circ\alpha) \in W_{\mathcal{B},0}$$

We define $A := \theta\circ H_{2,2}\circ H_{2,1}\circ H_1$.

Define

$$a := ((f, I_f, \alpha), (f\circ(A\circ\psi)^{-1}, I_{f\circ(A\circ\psi)^{-1}}, \beta^{-1}\circ\alpha), \psi\circ A) \tag{6.15}$$

$$\rho((f, I_f, \alpha), \beta) := a \tag{6.16}$$

The mapping ρ is a morphism: the automorphism ψ depends algebraically on (f, I_f, α) and β. In the slicing procedure we have seen that: if $\mathcal{B}$ is of type $A.1$, then A depends algebraically on (f, I_f, α), ψ; if $\mathcal{B}$ is of type $A.2$, then A depends algebraically on (f, I_f, α), ψ, (λ, μ).

PROPOSITION 6.26. *The morphism $\sigma_{\mathcal{B}} := t_{\mathcal{B}}\circ\rho$ is a group action and ρ is an isomorphism between the groupoid associated with the action and $W_{\mathcal{B},\bullet}$.*

PROOF. The following claim implies the proposition: given $g_1, g_2 \in \mathbf{G}_{\mathcal{B}}$ and $(f, I_f, \alpha) \in W_{\mathcal{B},\bullet}$

$$\rho(g_1, (f, I, \alpha)) \centerdot \rho(g_2, t_{\mathcal{B}}(\rho(g_1, (f, I, \alpha)))) = \rho(g_1 g_2, (f, I, \alpha))$$

We prove the claim: consider $g_1, g_2 \in \mathbf{G}_{\mathcal{B}}$ and $(f, I_f, \alpha) \in W_{\mathcal{B},\bullet}$. Let $g_i = (\beta_i, (\lambda_i, \mu_i))$; let $H_i(x,y) = (\lambda_i x, \mu_i y)$ for $i \leq 2$; then $g_1 g_2 = (\beta_1\circ\beta_2, (\lambda_1\lambda_2, \mu_1\mu_2))$; let $H_3(x,y) = (\lambda_1\lambda_2 x, \mu_1\mu_2 y)$.

Let $a_1 = \rho((f, I_f, \alpha), g_1)$. Because of the definition of ρ the arrow a_1 is of the form:

$$a_1 = ((f, I_f, \alpha), (f\circ(A_1\circ\psi_1)^{-1}, I_{f\circ(A_1\circ\psi_1)^{-1}}, \beta_1^{-1}\circ\alpha), A_1\circ\psi_1)$$

Let $g := f\circ(A_1\circ\psi_1)^{-1}$, let $a_2 := \rho((g, I_g, \beta_1^{-1}\circ\alpha), g_2)$. The arrow a_2 is of the form:

$$a_2 = ((g, I_g, \beta_1^{-1}\circ\alpha), (g\circ(A_2\circ\psi_2)^{-1}, I_{g\circ(A_2\circ\psi_2)^{-1}}, \beta_2^{-1}\circ\beta_1^{-1}\circ\alpha), A_2\circ\psi_2)$$

Let $a_3 = \rho((f, I_f, \alpha), g_1 g_2)$, the arrow a_3 is of the form:

$$a_3 = ((f, I_f, \alpha), (f\circ(A_3\circ\psi_3)^{-1}, I_{f\circ(A_3\circ\psi_3)^{-1}}, (\beta_1\circ\beta_2)^{-1}\circ\alpha), A_3\circ\psi_3)$$

We have to show that $a_1 \centerdot a_2 = a_3$. As

$$a_1 \centerdot a_2 = ((f, I_f, \alpha), (g\circ(A_2\circ\psi_2)^{-1}, I_{g\circ(A_2\circ\psi_2)^{-1}}, \beta_2^{-1}\circ\beta_1^{-1}\circ\alpha), A_2\circ\psi_2\circ A_1\circ\psi_1)$$

we see that the only thing to check is that $A_3\psi_3 = A_2\circ\psi_2\circ A_1\circ\psi_1$ There is a unique expression of $A_2\circ\psi_2\circ A_1\circ\psi_1$ as $A_3'\circ\psi_3'$ with A_3' affine and $\psi_3' \in G$. As $f\circ\psi_3'$ has behaviour graph isomorphic to $\mathcal{B}$ we deduce that ψ_3' is determined by the composable triple associated with (f, I_f, ψ_3'). This composable triple is determined by the image of the root by the automorphism of $Aut_{rm}(\mathcal{B})$ associated with it. As this image is $\beta_1\circ\beta_2(w)$, the same than the image of the root by the automorphism of $Aut_{rm}(\mathcal{B})$ associated with the composable triple corresponding to (f, I_f, ψ_3) we deduce that $\psi_3 = \psi_3'$.

If $\mathcal{B}$ is of case $A.1$ then we have seen that given any automorphism ψ such that $f \circ \psi^{-1}$ has behaviour graph $\mathcal{B}$, then there is a unique A, affine, such that $f \circ (A \circ \psi)^{-1}$ can be the first component of an element of $W_{\mathcal{B},0}$. Then, in this case, as $\psi_3 = \psi_3'$, we conclude that $A_3 = A_3'$ and we are done.

If $\mathcal{B}$ is of case $A.2$ then we have seen that, given any automorphism ψ such that $f \circ \psi^{-1}$ has behaviour graph $\mathcal{B}$, there exist a unique F of the form $F(x,y) = (x, y + bx)$ such that $f \circ (F \circ \psi)^{-1}$ can be the first component of an element of W_0^2. Let F_3 be the F that corresponds to $\psi_3 = \psi_3'$. Decompose A_3 and A_3' as $A_3 = \theta_3 \circ H_3 \circ F_3$ and $A_3' = \theta_3' \circ H_3' \circ F_3$, where H_3 and H_3' are linear and θ_3, θ_3' are translations. It is easy to see that both H_3 and H_3' must leave fixed the points $(0:1:0)$ and $(1:0:0)$, and hence be *diagonal* linear isomorphisms. We want to show that $H_3 = H_3'$. Consider the unique decomposition of H_3 (resp. H_3') as $H_3 = H_{3,2} \circ H_{3,1}$ (resp. $H_3' = H_{3,2}' \circ H_{3,1}'$), obtained by imposing that $H_{3,1}$ (resp. $H_{3,1}'$) is the multiplication of both coordinates by the same factor and that $H_{3,2}$ (resp. $H_{3,2}'$) belongs to $\mathbf{H}_{\mathcal{B}}^1$. As both $f \circ (\theta_3 \circ H_{3,2} \circ H_{3,1} \circ F_3 \circ \psi_3)^{-1}$ and $f \circ (\theta_3' \circ H_{3,2}' \circ H_{3,1}' \circ F_3 \circ \psi_3)^{-1}$ are the first components of elements of $W_{\mathcal{B},0}$ we deduce easily that both $f \circ (H_{3,1} \circ F_3 \circ \psi_3)^{-1}$ and $f \circ (H_{3,1}' \circ F_3 \circ \psi_3)^{-1}$ are the first components of elements of W_0^3. We know that given (f, I_f, α) and ψ there is only one couple $F_3, H_{3,1}$ of linear isomorphisms such that $f \circ (H_{3,1} \circ F_3 \circ \psi_3)^{-1}$ is the first component of an elements of W_0^3. This shows that $H_{3,1} = H_{3,1}'$. It remains to show that $H_{3,2} = H_{3,2}'$. For $i \in \{1,2\}$ factorise A_i as $H_i \circ F_i$ where $F_i(x,y) = (x, y + d_i x)$ and H_i is a diagonal linear isomorphism. A computation shows that $H_3' = H_2 \circ H_1$. Factorise each H_i as $H_{i,2} \circ H_{i,1}$ where $H_{i,1}$ is of the form $H_{i,1}(x,y) = (\gamma_i x, \gamma_i y)$ and $H_{i,2}(x,y) = (\lambda_i x, \mu_i y)$ belongs to $\mathbf{H}_{\mathcal{B}}^1$. The chain of equalities

$$H_{3,2}' \circ H_{3,1}' = H_3' = H_2 \circ H_1 = H_{2,2} \circ H_{2,1} \circ H_{1,2} \circ H_{1,1} = H_{2,2} \circ H_{1,2} \circ H_{2,1} \circ H_{1,1}$$

and the fact that each diagonal isomorphism decomposes *uniquely* in an isomorphism that takes (x,y) to $(\gamma x, \gamma y)$, for certain γ, and an element of $\mathbf{H}_{\mathcal{B}}^1$ implies that $H_{3,2}' = H_{2,2} \circ H_{1,2}$. As $H_{3,2}$ is by definition $H_{2,2} \circ H_{1,2}$ we have shown that $H_{3,2} = H_{3,2}'$ as we wanted. We would finish if we show that $\theta_3 = \theta_3'$. This follows from the fact that both $f \circ (\theta_3 \circ H_{3,2} \circ H_{3,1} \circ F_3 \circ \psi_3)^{-1}$ and $f \circ (\theta_3' \circ H_{3,2} \circ H_{3,1} \circ F_3 \circ \psi_3)^{-1}$ are the first components of elements of $W_{\mathcal{B},0}$. □

A consequence of the above proposition is that to construct a geometric quotient of $W_{\mathcal{B},\bullet}$ we just have to construct a quotient of $W_{\mathcal{B},0}$ by the group action $\sigma_{\mathcal{B}} : \mathbf{G}_{\mathcal{B}} \to W_{\mathcal{B},0}$. The subgroup $\mathbf{H}_{\mathcal{B}}^1 < \mathbf{G}_{\mathcal{B}}$ is normal. We will construct the geometric quotient in two steps: first we will quotient by $\mathbf{H}_{\mathcal{B}}^1$, and after by $\mathbf{G}_{\mathcal{B}}/\mathbf{H}_{\mathcal{B}}^1$.

PROPOSITION 6.27. *There exists an affine geometric quotient for the restricted action*

$$\sigma_{\mathcal{B}} : \mathbf{H}_{\mathcal{B}}^1 \times W_{\mathcal{B},0} \to W_{\mathcal{B},0}$$

PROOF. Let $(f, I_f, \leq_a) \in W_{\mathcal{B},0}$, let C be the curve defined by f, denote by $\tilde{C}$ and $\tilde{L}$ the strict transforms of C and L by the blowing up of $\mathbb{P}^2$ at $(0:1:0)$, and by E the exceptional divisor. The graph $\mathcal{B}$ determines the multiplicity m of C at $(0:1:0)$, and the intersection multiplicity $n := I_{\tilde{L} \cap E}(\tilde{C}, E)$. We have now that if $f(x,y) = \sum_{i+j \leq n} a_{i,j} x^i, y^j$ then $a_{m-l,n-m} \neq 0$. Consider the slice $W_{\mathcal{B},0}' \subset W_{\mathcal{B},0}$ given by the closed subset defined by the equation $a_{m-l,n-m} = 1$. Using Lemma 5.15 it is not difficult to reduce the proof to the construction of a geometric quotient of the groupoid $W'_{\mathcal{B},\bullet}$ obtained by restriction to $W'_{\mathcal{B},0}$. The set

$$\mathbf{H}_{\mathcal{B}} := \{(\lambda,\mu) \in \mathbf{H}^1_{\mathcal{B}} : \lambda^{m-l}\mu^{n-m} = 1\} \tag{6.17}$$

with pairwise multiplication is a group. As $(m-l, n-m)$ and $(s, -s+1)$ are linearly independent we deduce that it is finite. The groupoid $W'_{\mathcal{B},\bullet}$ is the groupoid associated by the finite group action

$$\rho_{\mathcal{B}} : \mathbf{H}_{\mathcal{B}} \times W'_{\mathcal{B},0} \to W'_{\mathcal{B},0} \tag{6.18}$$

obtained by restriction of $\sigma_{\mathcal{B}}$ As geometric quotients exist for finite group actions over affine variety this proves the proposition. □

PROPOSITION 6.28. *There exist a geometric quotient for* $W_{\mathcal{B},\bullet}$

PROOF. As $\mathbf{H}^1_{\mathcal{B}}$ is a normal subgroup of $\mathbf{G}_{\mathcal{B}}$ and $\mathbf{G}_{\mathcal{B}}/\mathbf{H}^1_{\mathcal{B}} \cong Aut_{rm}(\mathcal{B})$ we have an action

$$\sigma'_{\mathcal{B}} : Aut_{rm}(\mathcal{B}) \times (W_{\mathcal{B},0}/\mathbf{H}^1_{\mathcal{B}}) \to W_{\mathcal{B},0}/\mathbf{H}^1_{\mathcal{B}} \tag{6.19}$$

The existence of a geometric quotient for $W_{\mathcal{B},\bullet}$ is reduced to the existence of a geometric quotient for this group action. As $Aut_{rm}(\mathcal{B})$ is finite such quotient exists. □

Case A.3. Assume that there exists only one vertex of the first generation on $\mathcal{B}$. The first step of the slicing procedure is common for cases A.1, A.2 and A.3 and it consists here in fixing the unique point at infinity to be $(0:1:0)$.

LEMMA 6.29. *Let r be the number predicted in Lemma 6.8. Let $(f, I, \alpha) \in W^1_0$ and ψ, a triangular automorphism with graph $\mathcal{A}_n$ with $n \le r$. Then $f\circ\psi^{-1}$ has behaviour graph $\mathcal{B}$ and, moreover, for any isomorphism $\xi : \mathcal{B}_{f\circ\psi^{-1}} \to \mathcal{B}$ the triple*

$$((f, I_f, \alpha), (f\circ\psi^{-1}, I_{f\circ\psi^{-1}}, \xi), \psi)$$

is an arrow of W^1_1.

PROOF. The first assertion follows easily by taking into account Lemma 6.8 For the second notice that the slice W^1_0 has been constructed by fixing the point at infinity to be $(0:1:0)$ and this is not affected by the composition of f with ψ. □

Consider an arrow $((f, I_f, \alpha), (f\circ\phi^{-1}, I_{f\circ\phi^{-1}}, \xi), \phi) \in W^1_1$, as $\mathcal{B}$ is of minimal degree and the unique point at infinity of f is $(0:1:0)$ then ϕ admits a unique decomposition as $\phi = \theta\circ H\circ\psi$, where ψ is triangular, H is linear and θ is a translation. A computation shows that then ϕ admits a unique decomposition as $\phi = \theta_2\circ\varphi\circ\theta_1\circ H\circ\psi$ where the involved automorphisms are of the following form:

$$\begin{array}{rcl} \psi(x,y) & = & (x, y + \sum_{k>r}^{r} a_k x^k) \\ H(x,y) & = & (\lambda x, \mu y) \\ \theta_1(x,y) & = & (x+a, y) \\ \varphi(x,y) & = & (x, y + \sum_{k=1}^{r} a_k x^k) \\ \theta_2(x,y) & = & (x, y+b) \end{array} \tag{6.20}$$

There exists vertices $w, v_1, ..., v_{r+1}$ such that w is the root, $p(v_1) = w$, $p(v_2) = \{w, v_1\}$ and $p(v_i) = \{v_1, v_{i-1}\}$ if $i > 2$, any other vertex v of $\mathcal{B}$ satisfies $v <_p v_{r+1}$; moreover, there exists an arrow a attached to the edge conecting v_r with v_{r+1} (it is easy to see that if these vertices would not exist, either the number r introduced in Lemma 6.8 would be different or $\mathcal{B}$ would not be of minimal degree). Let $w, v_1, ..., v_n$ be the vertices of $\mathcal{B}$ ordered decreasingly by $\le_t$. We will say that a vertex v is of the *extended first floor* of $\mathcal{B}$ if it is either of the first floor or $p(v)$ is of the first floor. Let v be a free vertex of the extended first floor of $\mathcal{B}$ which is not in $\{w, v_1, ..., v_r\}$; let $\{w, v_{i_1}, ..., v_{i_j} = v\}$ be the set of vertices greater or equal than v by $\leq_p$; we define $d(v)$, *the distance of v to the second floor* as follows: if v_{i_h} is the smallest fixed vertex of $\{w, v_{i_1}, ..., v_{i_j} = v\}$ then define $d(v) := 2h - j$. Let $\mathcal{V}_1$ be the set of free vertices of the extended first floor of $\mathcal{B}$; define

$$D(\mathcal{B}) := min(\{d(v) : v \in \mathcal{V}_1\} \cup \{r\})$$

We will consider several types of graphs:

Type 1: There is a free vertex v_k with $p(v_k)$ free, $p(p(v_k))$ fixed and $p(p(v_k)) \neq v_{r+1}$. We choose v_k to be the maximal vertex of $\mathcal{B}$ respect $\leq_t$ with this property. Let $w, v_{i_1}, ..., v_{i_s} = v_k$ be the vertices greater or equal than v_k by $\leq_p$. If $D(\mathcal{B}) < r$ consider the vertex v_l, defined to be, among the free vertices of the extended first floor such that $d(v_l) = D(\mathcal{B})$, the biggest respect to $\leq_t$; let $v_{j_1}, ..., v_{j_t} = v_l$ be the vertices greater or equal than v_l by $\leq_p$.

Type 2: There is no vertex with the properties of v_k in last type 1; there exist two free vertices v, v' such that $p(v) \neq p(v')$, $p(v)$ and $p(v')$ are free and $p(p(v)) = p(p(v')) = v_{r+1}$. Note that $d(v) = d(v') = r - 1$. Let v_l be, among the free vertices of the extended first floor such that $d(v) = D(\mathcal{B})$, the biggest respect to $\leq_t$; let $w, v_{j_1}, ..., v_{j_t} = v_l$, as in type 1, note that $\{w, ..., v_{j_{r+1}}\} = \{w, v_1, ..., v_{r+1}\}$, and that v_{r+1} is the last fixed vertex of $w, ..., v_{j_t}$; hence $t = 2(r+1) - D(\mathcal{B})$. By the defining property of this type, there should exist a free vertex v_k such that $p(v_k) \neq v_{j_{r+2}}$ and $p(p(v_k)) = v_{r+1}$. Let $w, v_{i_1}, ..., v_{i_s} = v_k$ be the vertices greater or equal than v_k by $\leq_p$.

Types 1.1 and 2.1: Suppose that $\mathcal{B}$ is of type 1 (resp. 2), we will say that $\mathcal{B}$ is of type 1.1 (resp. 2.1) if there exist a free vertex $v_m \in \mathcal{B}$ such that $p(v_m)$ is free of the second floor. Note that in this case $D(\mathcal{B}) = 0$. Choose v_m to be the maximal with its property. Let $w, v_{h_1}, ..., v_{h_u} = v_m$ be the vertices greater or equal than v_m by $\leq_p$.

Types 2.1 and 2.2: We say that $\mathcal{B}$ is of type 1.2 (resp. 2.2) if it is of type 1 (resp. 2.) and not of type 1.1 (resp. 2.1).

Type 3: There exists a free vertex v such that $p(v)$ is free and $\mathcal{B}$ is not of one of the types above.

Type 4: There is no free vertex v such that $p(v)$ is free.

Given $\mathcal{B}$ of any of the types before, consider $(I, \alpha) \in V_{\mathcal{B}, \leq_a}$, for any vertex $v_i \in \mathcal{B}$ denote by x_i (resp. E_i) the point of I associated with v_i (resp. the exceptional divisor of the blowing up at x_i).

Definition 6.30. Define $Z_{\mathcal{B}} \subset V_{\mathcal{B}, \leq_a}$ to be the set of sequences $\{x_1, ..., x_n\} \in V_{\mathcal{B}, \leq_a}$ such that

(1) $x_1 = (0 : 1 : 0)$

(2) If $\mathcal{B}$ is of type 1 or 2, then the coordinates of $x_{i_{s-1}}$ and x_k respect to the canonical charts of $E_{i_{s-2}}$ and $E_{i_{s-1}}$ are respectively $x(x_{i_{s-1}}) = 1$ and $x(x_k) = 0$.

(3) If $\mathcal{B}$ is of type 1 or 2 and $D(\mathcal{B}) < r$, then the coordinates of $x_{j_{t-r+D(\mathcal{B})+1}}, ...,$ x_{j_t} respect to the canonical charts of $E_{i_{s-2}}$, $E_{i_{s-1}}$, $E_{j_{t-r+D(\mathcal{B})}}, ..., E_{j_{t-1}}$ are respectively $x(x_{j_{t-r+D(\mathcal{B})+1}}) = ... = x(x_{j_t}) = 0$.

(4) If $\mathcal{B}$ is of type 1.1 or 2.1 then the coordinate of x_m respect to the canonical chart of $E_{h_{u-1}}$ is $x(x_m) = 0$.

STEP 2. Define $W_0^2 \subset W_0^1$ by the formula $W_0^2 := \tau^{-1}_{\mathcal{B},\leq_a}(Z_{\mathcal{B}})$.

It is clear that if A is an affine automorphism and (f, I_f, α) is any ordered polynomial there is a canonical isomorphism between $\mathcal{B}_f$ and $\mathcal{B}_{f\circ A^{-1}}$. Moreover, if $\mathcal{B}$ is of case A.3 and φ is triangular of degree bouded by r, then, if we compute $\mathcal{B}_{f\circ\varphi^{-1}}$ starting from $\mathcal{B}_f$ and using composable triples, we see that there is also a canonical bijection between $\mathcal{B}_f$ and $\mathcal{B}_{f\circ\varphi^{-1}}$. Given an ordered polynomial $(f, I_f, \alpha) \in W_0^1$ and an automorphism σ that is a composition of affine and triangular automorphisms of degree bounded by r, we will identify by the canonical bijection $\mathcal{B}_f$ and $\mathcal{B}_{f\circ\sigma^{-1}}$ and obtain in this way an ordered polynomial $(f\circ\sigma^{-1}, I_{f\circ\sigma^{-1}}, \alpha)$.

We treat simultaneously the cases of graphs of type 1 and 2. Suppose that $(f, I_f, \alpha) \in W_0^1$. Lemma 6.17 shows that we can choose a unique H_1 of the form $H_1(x, y) = (\alpha x, \alpha y)$ such that the coordinate of the point $x_{i_{s-1}}$ of $I_{f\circ H_1^{-1}}$ respect to the canonical chart of $E_{i_{s-2}}$ is $x(x_{i_{s-1}}) = 1$; it also proves that, moreover, composition with automorphisms of the form of θ_1, φ, or θ_2 (see 6.20) does not move the coordinate of $x_{i_{s-1}}$. We define $W_0^{1,1} \subset W_0^1$ to be the set of $(f, I_f, \leq_a)$ such that the coordinate of the point $x_{i_{s-1}}$ of I_f respect to the canonical chart of $E_{i_{s-2}}$ is 1. Define

$$\mathbf{H}^1_{\mathcal{B}} := \{(\lambda, \mu) : \lambda^{s-2}\mu^{3-s} = 1\} \tag{6.21}$$

Consider $a := \big(((f, I_f, \alpha), (g, I_g, \beta)), \phi\big) \in W_1^2$. If we decompose ϕ as

$$\phi = \theta_2 \circ \varphi \circ \theta_1 \circ H_2 \circ H_1 \circ \psi$$

where H_1 is the automorphism chosen above, H_2 is a diagonal linear automorphism and θ_1, θ_2 and φ are like in formulae 6.20. Then H_1 is uniquely determined by (f, I_f, α), ψ, and depends algebraically on them. Moreover, by Lemma 6.17, if $H_2(x, y) = (\lambda x, \mu y)$, then (λ, μ) belongs to $\mathbf{H}^1_{\mathcal{B}}$.

Suppose that $\mathcal{B}$ is of type 1. Let $(f, I_f, \lambda) \in W_0^{1,1}$, Lemma 6.17 shows that:

(1) We can choose a unique number a such that, if $\theta_1(x, y) = (x + a, y)$, then the coordinate of the point x_k of $I_{f\circ\theta_1^{-1}}$ respect to the canonical chart of $E_{i_{s-1}}$ is 0.
(2) If $D(\mathcal{B}) < r$, then we can choose a unique number a_r such that, if $\varphi_0(x, y) = (x, y + a_r x^r)$, then the coordinate of the point $x_{j_{t-r+D(\mathcal{B})+1}}$ of $I_{f\circ(\varphi_0\circ\theta_1)^{-1}}$ respect to the canonical chart of $E_{j_{t-r+D(\mathcal{B})}}$ is 0.
(3) The lifting of φ_0 leaves $E_{i_{s-1}}$ fixed. Consequently the coordinate of the point x_k of $I_{f\circ(\varphi_0\circ\theta_1)^{-1}}$ respect to the canonical chart of $E_{i_{s-1}}$ is 0.

Suppose that $\mathcal{B}$ is of type 2. Consider $(f, I_f, \alpha) \in W_0^{1,1}$. Consider automorphisms $\theta_1(x, y) = (x + a, y)$ and $\varphi_0(x, y) = (x, y + a_r x^r)$. Lemma 6.17 states that the restrictions of the lifting of each automorphism to each of the exceptional divisors $E_{i_{s-1}}$ and $E_{j_{t-r+D(\mathcal{B})}}$, respect to the corresponding canonical charts are of the form

$$\theta_{1|E_{i_{s-1}}}(x) = x + A_{1,1}a \quad \theta_{1|E_{j_{t-r+D(\mathcal{B})}}}(x) = x + A_{1,2}a$$

$$\varphi_{0|E_{i_{s-1}}}(x) = x + A_{2,1}a \quad \varphi_{0|E_{j_{t-r+D(\mathcal{B})}}}(x) = x + A_{2,2}a$$

The computations needed to prove Lemma 6.17, specialized to this case, show also that the determinant of the square matrix $(A_{i,j})$ is non zero. To force the coordinates of the points x_{i_s} and $x_{j_{t-r+D(\mathcal{B})+1}}$ of $I_{f\circ(\varphi_0\circ\theta_1)^{-1}}$ respect to the corresponding canonical charts to be both 0, the parameters a and a_r should be a solution of a linear system that have $(A_{i,j})$ as fundamental matrix. This implies that there are unique θ_1 and φ_0 with the desired property.

Now we resume the simultaneous treatment of types 1 and 2. We define $W_0^{1,2} \subset W_0^{1,1}$ to be the set of $(f, I_f, \leq_a)$ such that the coordinates of the points x_{i_s} and $x_{j_{t-r+D(\mathcal{B})+1}}$ of I_f respect to the corresponding canonical charts are both 0.

Consider $a := \big(((f, I_f, \alpha), (g, I_g, \beta)), \phi\big) \in W_1^2$. Decompose ϕ as

$$\phi = \theta_2 \circ \varphi' \circ \varphi_0 \circ \theta_1 \circ H_2 \circ H_1 \circ \psi$$

where

$$\varphi_0(x, y) = (x, y + a^r x^r) \quad \varphi'(x, y) = (x, y + \sum_{i=1}^{r-1} a_i x^i)$$

then θ_1 and φ_0 are uniquely determined by (f, I_f, α), ψ, H_2 and depend algebraically on them.

Let $(f, I_f, \lambda) \in W_0^{1,2}$, Lemma 6.17 shows that we can succesively choose unique a_{r-1},...,$a_{D(\mathcal{B})+1}$ such that, with an arbitrary choice of a_1,...$a_{D(\mathcal{B})}$, the coordinates of the points $x_{j_{t-r+D(\mathcal{B})+2}}, \ldots, x_{j_t}$ of $I_{f \circ (\varphi')^{-1}}$ are all of them 0.

Consider $a := \big(((f, I_f, \alpha), (g, I_g, \beta)), \phi\big) \in W_1^2$. Decompose ϕ as

$$\phi = \theta_2 \circ \varphi_2 \circ \varphi_1 \circ \varphi_0 \circ \theta_1 \circ H_2 \circ H_1 \circ \psi \tag{6.22}$$

where

$$\varphi_1(x, y) = (x, y + \sum_{i=D(\mathcal{B})+1}^{r-1} a_i x^i) \quad \varphi_2(x, y) = (x, y + \sum_{i=1}^{D(\mathcal{B})} a_i x^i)$$

and the rest of the automorphisms are as above. Then φ_1 is uniquely determined by (f, I_f, α), ψ, H_2 and depends algebraically on them.

If $\mathcal{B}$ is of type 1.2 or 2.2 then Lemma 6.17 implies that if $(f, I_f, \alpha) \in W_0^2$ then, if $\phi = \theta_2 \circ \varphi_2$ with θ_2 and φ_2 as above then $I_f = I_{f \circ \phi^{-1}}$.

Suppose that $\mathcal{B}$ is of type 1.1 or 2.1 and that $(f, I_f, \alpha) \in W_0^{1,1}$ is such that the coordinates of x_k, $x_{j_{t-r+D(\mathcal{B})+1}}, \ldots, x_{j_t}$ are the desired ones. Lemma 6.17 implies that there exists a unique θ_2 of the form of formulae 6.20 such that $(f \circ \theta_2^{-1}, I_{f \circ \theta_2^{-1}}, \alpha) \in W_0^2$.

Consider $a := \big(((f, I_f, \alpha), (g, I_g, \beta)), \phi\big) \in W_1^2$. Decompose ϕ as

$$\phi = \theta_2 \circ \varphi \circ \theta_1 \circ H_2 \circ H_1 \circ \psi$$

Then θ_2 is uniquely determined by (f, I_f, α), ψ, H_2 and depends algebraically on them.

If $\mathcal{B}$ is of type 3 or 4 then $W_0^2 := W_0^1$. Note that if $\mathcal{B}$ is of type 4 and $(f, I_f, \alpha) \in W_0^2$, then given any ϕ that admits a decomposition as $\phi = \theta_2 \circ \varphi \circ \theta_1$, with θ_2, ϕ and θ_1 like in formulae 6.20, we have that $I_f = I_{f \circ \phi^{-1}}$.

Before we continue with the slicing procedure we need the following discussion: let $(f, I, \alpha) \in W_0^1$, the only point at infinity of f is $p = (0 : 1 : 0)$; hence $I_p(C, L) = n$, the degree of $\mathcal{B}$. Let $m := \mathrm{mult}_p(C)$. Consider the blowing up process whose first center is p and whose parental system is isomorphic to the one formed by $\{w, v_1, ..., v_{r+1}\}$, the vertices of $\mathcal{A}_{r+1}$ bigger or equal than the fixed central vertex. Let E_i be the exceptional divisor corresponding to the vertex v_i, let $\tilde{C}_i$ be the strict transform of C by the i first blowing ups. The graph $\mathcal{B}$ determines the numbers

$l := I_{L\cap E_1}(\tilde{C}_1, E_1)$, $m' := \mathrm{mult}_{E_1\cap E_r}(\tilde{C}_r)$ and $l' := I_{E_r\cap E_{r+1}}(\tilde{C}_{r+1}, E_{r+1})$. An easy argument using Newton polygons shows that if $f(x,y) = \sum a_{i,j}x^iy^j$ then

$$
\begin{array}{lll}
& a_{i,j} = 0 & \text{if } \; j > n - m \\
& a_{i,j} = 0 & \text{if } \; i + rj > n \\
& a_{i,j} = 0 & \text{if } \; i + rj = n \quad \text{and} \quad i \neq n \\
& a_{n,0} \neq 0 & \\
(6.23) & a_{i,j} = 0 & \text{if } \; i + (r+1)j > 2n - m - m' \\
& a_{i,j} = 0 & \text{if } \; i + (r+1)j = 2n - m - m' \\
& & \quad \text{and } \; j < n - m - m' + l' \\
& a_{rm-(r-1)n+rm'-(r+1)l',n-m-m'+l'} \neq 0 & \\
& a_{m-l,n-m} \neq 0 &
\end{array}
$$

To abreviate formulae it will be convenient to define $(N, M) := (rm - (r-1)n + rm' - (r+1)l', n - m - m' + l')$ and $(N', M') := (m - l, n - m)$

STEP 3. If $\mathcal{B}$ is of type 1 or 2 define $W_0^3 = W_0^2$. Suppose that $\mathcal{B}$ is of type 3 or 4; define $W_0^3 \subset W_0^2$ to be the set of $(f, I_f, \alpha) \in W_0^2$ such that if $f(x,y) = \sum a_{i,j}x^iy^j$ then $a_{N',M'} = 1$. We reduce the quotienting of $W_\bullet^2$ to the quotienting of $W_\bullet^3$ exactly in the same way than in the previous cases.

Consider $a := \big(((f, I_f, \alpha), (g, I_g, \beta)), \phi\big) \in W_1^3$. Decompose ϕ as $\theta_2\circ\varphi\circ\theta_1\circ H\circ\psi$ where the involved automorphisms are like in formulae 6.20. Suppose that $f\circ\phi^{-1} = \sum_{i+j\leq n} b_{i,j}x^iy^j$; then a belongs to W_1^3 if and only if $b_{N',M'} = 1$. Taking into account equations 6.23, an easy computation shows that $b_{N',M'} = 1$ if and only if $\lambda^{N'}\lambda^{M'} = 1$. The set

$$
\mathbf{H}_{\mathcal{B}}^1 := \{(\lambda, \mu) : \lambda^{N'}\mu^{M'} = 1\} \tag{6.24}
$$

with pairwise multiplication is a group isomorphic to the direct product $\mu_D \times \mathbb{C}^*$, where $D = mcd(m - l, n - m)$ and μ_D denotes the group of D-th roots of unity. We have shown that $a \in W_0^3$ if and only if $(\lambda, \mu) \in \mathbf{H}_{\mathcal{B}}^1$.

STEP 4. We define $W_0^4 \subset W_0^3$ by the equations

$$
a_{N',M'-1} = a_{N'+1,M'-1} = \ldots = a_{N'+D(\mathcal{B}),M'-1} = 0
$$

if $\mathcal{B}$ is of type 1.2 or 2.2, and by the equations

$$
a_{n-1,0} = a_{N,M-1} = a_{N+1,M-1} = \ldots = a_{N+r,M-1} = 0
$$

if $\mathcal{B}$ is of type 3 or 4. If $\mathcal{B}$ is of types 1.1 or 2.1 then $W_0^4 := W_0^3$.

We explain why the quotienting of $W_\bullet^3$ can be reduced to the quotienting of $W_\bullet^4$ in the case of a graph of types 3 or 4. For the other types the reasoning is analogous. Consider $(f, I_f, \alpha) \in W_0^3$ with $f(x,y) = \sum a_{i,j}x^iy^j$, take $\sigma(x,y) = (x, y + \frac{\sum_{i=0}^r a_{N+i,M-1}x^i}{Ma_{N,M}})$. Using the formulae 6.23 one shows easily that if $f\circ\sigma^{-1} = \sum b_{i,j}x^iy^j$ then $b_{N,M-1} = b_{N+1,M-1} = \ldots = b_{N+r,M-1} = 0$. Take $\theta(x,y) = (x - \frac{b_{n-1,0}}{nb_{n,0}}, y)$; again using formulae 6.23 we show that $(f\circ(\theta\circ\sigma)^{-1}, I_{f\circ(\theta\circ\sigma)^{-1}}, \alpha) \in W_0^4$.

REMARK 6.31. The reason why we impose the equations $a_{N+i,M-1} = 0$ for $i \leq +D(\mathcal{B})$ in types 3 and 4, instead of $a_{N'+i,M'-1} = 0$, like in other types is that composition with θ could affect the coefficients $b_{N'+i,M'-1}$ and destroy the equations $b_{N'+i,M'-1} = 0$, and does not affect the coefficients $b_{N+i,M-1}$, preserving then the equations $b_{N+i,M-1} = 0$.

A computation shows that there are unique θ_1, θ_2, φ like in formulae 6.20 such that $\theta\circ\sigma = \theta_2\circ\varphi\circ\theta_1$.

Consider $a := (((f, I_f, \alpha), (g, I_g, \beta)), \phi) \in W_1^4$. Decompose ϕ as

$$\theta_2 \circ \varphi \circ \theta_1 \circ H_2 \circ H_1 \circ \psi$$

the last discussion implies that θ_1, θ_2 and φ are uniquely determined by (f, I_f, α), ψ, H_2.

We define the groupoid that we are going to be able to quotient:

$$W_{\mathcal{B},\bullet} := W_\bullet^4 \tag{6.25}$$

We sumarize some statements obtained in the slicing procedure:

LEMMA 6.32. *If $a = ((f, I_f, \alpha), (g, I_g, \beta), \phi)$ is an arrow of $W_{\mathcal{B},1}$ and ϕ is decomposed as $\theta_2 \circ \varphi \circ \theta_1 \circ H_2 \circ H_1 \circ \psi$ as above, then H_1 is uniquely determined by (f, I_f, α), ψ and depends algebraically on them; the automorphism H_2 belongs to $\mathbf{H}_{\mathcal{B}}^1$; finally, φ, θ_1 and θ_2 are uniquely determined and depend algebraically on (f, I_f, α), ψ, H_2.*

Consider any $(f, I_f, \alpha) \in W_{\mathcal{B},0}$. If $\mathcal{B}$ is of type 1.2 or 2.2 consider any $\sigma \in Aut(\mathbb{C}^2)$ of the form $\sigma(x, y) = (x, y + \sum_{i=0}^{D(\mathcal{B})})$; if $\mathcal{B}$ is of type 4 consider any $\sigma \in Aut(\mathbb{C}^2)$ of the form $\sigma(x, y) = (x + a, y + \sum_{i=0}^{r})$. Then $I_f = I_{f \circ \sigma^{-1}}$.

As in case A.1 we are going to prove that the groupoid $W_{\mathcal{B},\bullet}$ is actually a group action. Define the group $\mathbf{G}_{\mathcal{B}}$ to be the direct product

$$\mathbf{G}_{\mathcal{B}} := Aut_{rm}(\mathcal{B}^m) \times \mathbf{H}_{\mathcal{B}}^1$$

We define a mapping

$$\rho : \mathbf{G}_{\mathcal{B}} \times W_{\mathcal{B},0} \to W_{\mathcal{B},1}$$

in the following way: consider $(f, I_f, \alpha) \in W_{\mathcal{B},0}$ and $(\beta, (\lambda, \mu)) \in \mathbf{G}_{\mathcal{B}}$. Let $\Xi(\beta) = (\gamma, \mathcal{B}, (S, p), (S_0, p_0))$. Define $(S_f, p_f) := \alpha^{-1}(S_0, p_0)$. Consider the composable triple $\mathcal{B}_f, (S, p), (S_f, p_f)$. Following the reasoning of the proof of Lemma 4.7 we see that there exists a triangular automorphism $\psi \in G$ such that the composable triple associated with (f, I_f, ψ) is $\mathcal{B}_f, (S, p), (S_f, p_f)$. A straightforward computation shows the following relation between automorphisms and their indetermination points: given $\psi(x, y) = (x, y + \sum_{i=2}^{s} c_i x^i)$ then, if $x_1, ..., x_{2s-1}$ is the set of infinitely near points given by the indeterminacy of ψ, then the coefficient c_i is determined by the location of the infinitely near point x_{2s-i+1}. As $x_{s+1}, ..., x_{2s-r}$ belong to $\mathcal{B}_f$ the coefficients $c_{r+1}, ..., c_s$ are uniquely determined by (f, I_f, α) and $\mathcal{B}_f, (S_f, p_f), (S_f, p_f)$, and hence by (f, I_f, α) and β. In the slicing procedure we have shown that there exist unique $c_2, ..., c_r$ such that an arrow of the form

$$a := ((f, I_f, \alpha), f \circ (A \circ \psi)^{-1}, I_{f \circ (f \circ A \circ \psi)^{-1}}, \xi), A \circ \psi) \tag{6.26}$$

(where A is certain affine isomorphism) can be in $W_{\mathcal{B},1}$. Choose ψ by letting the coefficients $c_2, ..., c_r$ be such that this property is satisfied. Then ψ depends algebraically on $(f, I_f, \alpha), \beta$. Decompose A as $A = A_2 \circ H_1$, where $H_1(x, y) = (\lambda' x, \mu' y)$ and $A_2(x, y) = (x + d, bx + y + e)$; in the slicing procedure we have shown that the condition $a \in W_{\mathcal{B},1}$ is equivalent to the facts that $(\lambda', \mu') \in \mathbf{H}_{\mathcal{B}}^1$ and that b, d and e assume some values that depend algebraically on (f, I_f, α), ψ, H_1. Choose $H_1(x, y) = (\lambda c, \mu y)$ and b, d and e to be the values corresponding to (f, I_f, α), ψ and H_1 so that $a \in W_{\mathcal{B},1}$. Define $\phi := A_2 \circ H_1 \circ \psi$.

The graph $\mathcal{B}_{f \circ \phi^{-1}}^m$ can be identified with the result of letting the $\alpha^{-1}(\beta(v_r))$ become the root of $\mathcal{B}_f^m$ (where v_r is the root of $\mathcal{B}^m$). We note that

$$\beta^{-1} \circ \alpha : \mathcal{B}_{f \circ \phi}^m \to \mathcal{B}^m$$

is an isomorphism of rooted graphs. The arrow

$$a := \big((f, I_f, \alpha), (f \circ \phi^{-1}, I_{f\circ\phi^{-1}}, \beta^{-1} \circ \alpha), \phi\big)$$

is in $W^1_{\mathcal{B}}$. We define

$$\rho((f, I_f, \alpha), \beta) := a \tag{6.27}$$

where a is the arrow defined just above. We have already checked that ρ is morphism.

PROPOSITION 6.33. *The morphism $\sigma_{\mathcal{B}} := t_{\mathcal{B}} \circ \rho$ is a group action and ρ is an isomorphism between the groupoid associated with the action and $W_{\mathcal{B},\bullet}$.*

PROOF. The proof is similar to the proof of Proposition 6.26. □

A consequence of the above proposition is that to construct a geometric quotient of $W_{\mathcal{B},\bullet}$ we just have to construct a group quotient of $W_{\mathcal{B},0}$ by the group action $\sigma_{\mathcal{B}} : \mathbf{G}_{\mathcal{B}} \times W_{\mathcal{B},0} \to W_{\mathcal{B},0}$. The subgroup $\mathbf{H}^1_{\mathcal{B}} < \mathbf{G}_{\mathcal{B}}$ is a normal subgroup.

PROPOSITION 6.34. *There exists an affine geometric quotient for the restricted action $\sigma_{\mathcal{B}} : \mathbf{H}^1_{\mathcal{B}} \times W_{\mathcal{B},0} \to W_{\mathcal{B},0}$*

PROOF. The proof is identical to Proposition 6.27, but using the slice $W'_{\mathcal{B},0}$ defined by the equation $a_{n,0} = 1$. The finite group that will arise in the quotienting of the slice will be

$$\mathbf{H}_{\mathcal{B}} := \{(\lambda, \mu) : (\lambda, \mu) \in \mathbf{H}^1_{\mathcal{B}} \quad \text{and} \quad \lambda^n = 1\} \tag{6.28}$$

□

PROPOSITION 6.35. *There exist a geometric quotient for $W_{\mathcal{B},\bullet}$*

PROOF. As $\mathbf{H}^1_{\mathcal{B}}$ is a normal subgroup of $\mathbf{G}_{\mathcal{B}}$ and $\mathbf{G}_{\mathcal{B}}/\mathbf{H}^1_{\mathcal{B}} \cong Aut_{rm}(\mathcal{B}^m)$ all we need to do is to construct a quotient for the action

$$\sigma'_{\mathcal{B}} : Aut_{rm}(\mathcal{B}^m) \times (W_{\mathcal{B},0}/\mathbf{H}^1_{\mathcal{B}}) \to W_{\mathcal{B},0}/\mathbf{H}^1_{\mathcal{B}} \tag{6.29}$$

As $Aut_{rm}(\mathcal{B}^m)$ is a finite group we deduce that a geometric quotient exists. □

6.2.3. Case B. The graph $\mathcal{B}$ is a behaviour graph of minimal degree such that $\mathbf{CT}(\mathcal{B})$ only has the trivial element and $\mathcal{B}$ is not equivalent to the behaviour of a set of parallel lines.

In this case if $f \in X_{\mathcal{B},0}$ and $(f, \phi) \in X_{\mathcal{B},1}$ then ϕ must be linear. We will consider several subcases. As the kind of reasoning is quite similar to Case A, we will just indicate how to construct the slice in each of the cases, and which group action remains after the slicing.

Case B.1. The graph $\mathcal{B}$ has at least three vertices of the first generation. Given $(f, I_f, \alpha) \in X_{\mathcal{B},\leq_a}$ we denote by x_1, x_2, x_3 the points at infinity corresponding to v_1, v_2, v_3, the first three vertices of the first generation of $\mathcal{B}$. We will consider different types of graphs depending of the presence in $\mathcal{B}$ of the following types of vertices:

- A free vertex v_k such that if $\{w, v_{i_1}, ..., v_{i_r} = v_k\}$ are the points of $\mathcal{B}$ greater or equal than v_k, then $p(v_k) = \{v_{i_{r-1}}\}$ and the parental system restricted to $\{w, v_{i_1}, ..., v_{i_{r-1}}\}$ is isomorphic to the restriction of parental system of the graph $\mathcal{A}_{r-2}$ to the vertices greater or equal than the first free lateral vertex.

• If v_k appears, another free vertex v_l such that the points $\{w, v_{j_1}, ..., v_{j_s} = v_l\}$ of $\mathcal{B}$ greater or equal than v_l by the parental order, satisfy $p(v_l) = \{v_{j_{s-1}}\}$ and , in addition, one the two following properties:

(1) The parental system restricted to $\{w, v_{j_1}, ..., v_{j_{s-1}}\}$ is isomorphic to the restriction of parental system of the graph $\mathcal{A}_{s-2}$ to the vertices greater or equal than the first free lateral vertex. Moreover $v_{i_1} \neq v_{j_1}$.

(2) The parental system restricted to $\{w, v_{j_1}, ..., v_{j_{s-1}}\}$ is isomorphic to the restriction of parental system of the graph $\mathcal{A}_{n,m}$ for certain n, m to the vertices greater or equal than the first free lateral vertex of second floor. Moreover $v_{i_1} = v_{j_1}$.

Type 1: Both v_k and v_l exist in $\mathcal{B}$. For simplicity we assume the following: if v_l satisfies the first of the two possible properties then $v_{i_1} = v_1$ and $v_{j_1} = v_2$; if v_l satisfies the second property then $v_{j_1} = v_{j_2} = v_1$.

Type 2: The vertex v_k exists in $\mathcal{B}$, and there is no vertex with the properties of v_l. Assume for simplicity that $v_{i_1} = v_1$

Type 3: There are no vertices in $\mathcal{B}$ with the properties of v_k and v_l.

Given any $(I, \alpha) \in V_{\mathcal{B},\leq_a}$, for $h \leq r$ (resp. for $h \leq s$) we will denote by x_{i_h} and E_{i_h}, (resp. x_{j_h} and E_{j_h}) the infinitely near point associated with v_{i_h} (resp. v_{j_h}) and the exceptional divisor of the blowing up at x_{i_h} (resp. x_{j_h}).

Define $Z_{\mathcal{B}} \subset V_{\mathcal{B},\leq_a}$ to be the closed subset formed by sets of infinitely near points such that the following conditions are satisfied: $x_1 = (0:1:0)$, $x_2 = (1:0:0)$ and $x_3 = (1:1:0)$, and the coordinates of x_k and x_l (provided they exist) respect to the canonical charts of $E_{i_{r-1}}$ and $E_{j_{s-1}}$ are $x(x_k) = 0$ and $x(x_l) = 0$.

Define $W_0^1 := \tau_{\mathcal{B},\leq_a}^{-1}(Z_{\mathcal{B}})$. Let $(f, I_f, \alpha) \in W_0^1$. By Lemma 6.17, if $\mathcal{B}$ is of type 2 then for any translation of the form $\theta(x,y) = (x, y+e)$ we have that $I_{f\circ\theta^{-1}} = I_f$; if $\mathcal{B}$ is of type 3 and θ is *any* translation then $I_{f\circ\theta^{-1}} = I_f$.

Let $d := I_{p_1}(C, L)$ and $d' := I_{p_2}(C, L)$ (notice that d and d' are determined by $\mathcal{B}$). We define $W_{\mathcal{B},\bullet}$ to be the subset of triples $(f, I_f, \alpha) \in W_0^1$ such that if $f(x,y) = \sum a_{i,j} x^i y^j$ then

- If $\mathcal{B}$ is of type 1 $a_{d,n-d} = 1$.
- If $\mathcal{B}$ is of type 2 $a_{d,n-d} = 1$ and $a_{n-d',d'-1} = 0$
- If $\mathcal{B}$ is of type 3 $a_{d,n-d} = 1$, $a_{n-d',d'-1} = 0$ and $a_{d-1,n-d} = 0$.

It is easy to check that the groupoid $W_{\mathcal{B},\bullet}$ is such that both the source and the target map are finite and with analytic local sections. This implies that both the source and the target map are a covering morphism, and hence they are finite and flat. Under this condition it is known the existence of a geometric quotient for $W_{\mathcal{B},\bullet}$ (for a proof see Proposition 5.1 of [**22**]).

Let $\mathbf{H}_{\mathcal{B}}$ denote the group of n-th roots of unity. Consider the group action $\sigma : \mathbf{H}_{\mathcal{B}} \times W_{\mathcal{B},0} \to W_{\mathcal{B},0}$ defined by letting $\xi \in \mathbf{H}_{\mathcal{B}}$ act as the homotecy $H(x,y) = (\xi x, \xi y)$. As $\mathbf{H}_{\mathcal{B}}$ is a finite group and $W_{\mathcal{B},0}$ is affine there exists an affine geometric quotient Q for this group action. We define a group action $\rho : Aut(\mathcal{B}) \times Q \to Q$ as follows: consider $\beta \in Aut(\mathcal{B})$ and $[(f, I_f, \alpha)]$, a class of Q. Given any affine isomorphism H there is a canonical bijection γ between $I_{f\circ H^{-1}}$ and I_f. This defines an isomorphism $\gamma : \mathcal{B}_{f\circ H^{-1}} \to \mathcal{B}_f$. There exists a linear isomorphism H, that is unique up to multiplication by a constant, such that the triple $(f\circ H^{-1}, I_{f\circ H^{-1}}, \beta\circ\alpha\circ\gamma)$ belongs to W_0^1. Given H there exists a unique translation

θ such that $(f\circ(\theta\circ H)^{-1}, I_{f\circ(\theta\circ H)^{-1}}, \beta\circ\alpha\circ\gamma)$ is in $W_{\mathcal{B},0}$. We define

$$\rho(\beta, [(f, I_f, \alpha)]) := [(f\circ(\theta\circ H)^{-1}, I_{f\circ(\theta\circ H)^{-1}}, \beta\circ\alpha\circ\gamma)]$$

It is easy to check that the definition does not depend on the choice of H, and that ρ is a group action. As $Aut(\mathcal{B})$ is a finite group then there exists a geometric quotient Q' for the action ρ. Let $q : W_{\mathcal{B},0} \to Q'$ be the composition of the quotienting maps of the actions defined above. Using the techniques of section 7 of [**22**] it is possible to show that $q : W_{\mathcal{B},0} \to Q'$ is the geometric quotient of the groupoid $W_{\mathcal{B},\bullet}$. Notice that to check this assertion at the level of points is trivial. We have

LEMMA 6.36. *The groupoid $W_{\mathcal{B},\bullet}$ has an affine geometric quotient that is isomorphic to the result of the succesive quotient by the actions of $\mathbf{H}_{\mathcal{B}}$ on $W_{\mathcal{B},0}$ and of $Aut(\mathcal{B})$ on $W_{\mathcal{B},0}/\mathbf{H}_{\mathcal{B}}$.*

Case B.2. The graph $\mathcal{B}$ has exactly two vertices of the first generation and at least one of the second. This means that if f is any polynomial with behaviour $\mathcal{B}$ then the curve that it defines is tangent to the line at infinity in at least one of the points. Given an admissible ordering $\leq_a$ we denote by v_1 and v_2 the first and second points of the first generation. We will assume for simplicity that if only one of v_1, v_2 is greater by the parental order than a vertex of the second generation, then it is v_1. We will distinguish three types of graphs exactly like in Case B.1.

If $v_1, ..., v_n$ are the vertices of $\mathcal{B}$ ordered by the total ordering associated to $\leq_a$ and $(f, I_f, \alpha) \in V_{\mathcal{B},\leq_a}$ then we denote by x_i the point of I_f associated with v_i.

We define $Z_{\mathcal{B}}$ like in the case before but without the assumption $x_3 = (1 : 1 : 0)$ (it does not make sense in this case).

Define $W_0^1 := \tau^{-1}_{\mathcal{B},\leq_a}(Z_{\mathcal{B}})$. Let $(f, I_f, \alpha) \in W_0^1$. By Lemma 6.17, if $\mathcal{B}$ is of type 2 then for any translation of the form $\theta(x, y) = (x, y+e)$ we have that $I_{f\circ\theta^{-1}} = I_f$, and if $\mathcal{B}$ is of type 3 and θ is *any* translation then $I_{f\circ\theta^{-1}} = I_f$.

Let $d := I_{x_1}(C, L)$ (notice that d is determined by $\mathcal{B}$). We define $W_{\mathcal{B},\bullet}$ to be the subset formed by the ordered polynomials $(f, I_f, \alpha) \in W_0^1$ such that if $f(x, y) = \sum a_{i,j}x^iy^j$ then

- If $\mathcal{B}$ is of type 1 $a_{d,n-d} = 1$.
- If $\mathcal{B}$ is of type 2 $a_{d,n-d} = 1$ and $a_{n-d,d-1} = 0$
- If $\mathcal{B}$ is of type 3 $a_{d,n-d} = 1$, $a_{n-d,d-1} = 0$ and $a_{d-1,n-d} = 0$.

Consider the group (whose operation is pairwise multiplication)

$$\mathbf{H}^1_{\mathcal{B}} := \{(\lambda, \mu) : \lambda^d\mu^{n-d} = 1\}$$

$\mathbf{H}_{\mathcal{B}}$ is a group isomorphic to $\mathbb{C}^* \times \mu_D$ where $D = gcd(d, n-d)$. Define the set

$$\mathbf{G}_{\mathcal{B}} := \mathbf{H}^1_{\mathcal{B}} \times Aut(\mathcal{B})$$

we give to it structure of group defining the multiplication as follows: if β' does not permute the vertices of the first generation then

$$((\lambda, \mu), \beta) \centerdot ((\lambda', \mu'), \beta') := ((\lambda'\lambda, \mu'\mu), \beta\circ\beta')$$

if β' permutes the vertices of the first generation then

$$((\lambda, \mu), \beta) \centerdot ((\lambda', \mu'), \beta') := ((\lambda'\mu, \mu'\lambda), \beta\circ\beta')$$

We define the map:

$$\rho : \mathbf{G}_{\mathcal{B}} \times W_{\mathcal{B},0} \to W_{\mathcal{B},1}$$

in the following way: Consider $((\lambda,\mu),\beta) \in \mathbf{G}_{\mathcal{B}}$ and $(f, I_f, \alpha) \in W_{\mathcal{B},0}$. Let $H(x,y) := (\lambda x, \mu y)$. If β' does not permute the vertices of the first generation then define $\tau = id_{\mathbb{C}^2}$; if β' permutes the vertices of the first generation then define $\tau(x,y) := (y,x)$. Define

$$\rho\big(((\lambda,\mu),\beta),(f,I_f,\alpha)\big) := \left(f\circ(H\circ\tau)^{-1}, I_{f\circ(H\circ\tau)^{-1}}, \beta\circ\alpha\right)$$

It is easy to see that $t_{\mathcal{B}}\circ\rho : \mathbf{G}_{\mathcal{B}} \times W_{\mathcal{B},0} \to W_{\mathcal{B},1}$ is a group action and that ρ defines an isomorphism between the groupoid associated with the group action and $W_{\mathcal{B},\bullet}$. Hence to construct a geometric quotient of $W_{\mathcal{B},\bullet}$ we just have to construct a quotient of $W_{\mathcal{B},0}$ by the group action $\sigma_{\mathcal{B}} : \mathbf{G}_{\mathcal{B}} \to W_{\mathcal{B},0}$.

The subgroup $\mathbf{G}^1_{\mathcal{B}} < \mathbf{G}_{\mathcal{B}}$ formed by the pairs $((\lambda,\mu),\beta)$ such that β does not permute the vertices of the first generation is a normal subgroup of index 2. We can construct the geometric quotient in two steps: first we will quotient by $\mathbf{G}^1_{\mathcal{B}}$, and after by $\mathbf{G}_{\mathcal{B}}/\mathbf{G}^1_{\mathcal{B}} \cong S_2$ (where S_2 is the permutation group of two elements). The subgroup $\mathbf{H}^1_{\mathcal{B}} < \mathbf{G}^1_{\mathcal{B}}$ is a normal subgroup. We will construct the quotient by $\mathbf{G}^1_{\mathcal{B}}$ in two steps: first we will quotient by $\mathbf{H}^1_{\mathcal{B}}$, and after by $\mathbf{G}^1_{\mathcal{B}}/\mathbf{H}^1_{\mathcal{B}}$.

PROPOSITION 6.37. *There exists an affine geometric quotient for the restricted action* $\sigma_{\mathcal{B}} : \mathbf{H}^1_{\mathcal{B}} \times W_{\mathcal{B},0} \to W_{\mathcal{B},0}$

PROOF. Given (f, I_f, α) let C be the curve defined by f. Let E and $\tilde{C}$ be the exceptional divisor and the strict transform of C by the blowing up at x_1. Let $l := I_{E_1\cap L}(\tilde{C}_1, E_1)$ and $m := \mathrm{mult}_{x_1}(C)$, from now on the proof is identical to Proposition 6.27 but using the slice $W'_{\mathcal{B},0} \subset W_{\mathcal{B},0}$ defined by the equation $a_{m-l,n-m} = 1$. The finite group that will arise in the quotienting of the slice will be

$$\mathbf{H}_{\mathcal{B}} := \{(\lambda,\mu) \in \mathbf{H}^1_{\mathcal{B}} : \lambda^{m-l}\mu^{n-m} = 1\} \tag{6.30}$$

□

PROPOSITION 6.38. *There exists an affine geometric quotient of* $W_{\mathcal{B},\bullet}$

PROOF. The group $\mathbf{H}^1_{\mathcal{B}}$ is a normal subgroup of $\mathbf{G}^1_{\mathcal{B}}$ and $\mathbf{G}^1_{\mathcal{B}}/\mathbf{H}^1_{\mathcal{B}} \cong Aut^1(\mathcal{B})$, where $Aut^1(\mathcal{B})$ is the set of automorphisms that leave fixed the vertices of the first generation. We have an action

$$\sigma'_{\mathcal{B}} : Aut^1(\mathcal{B}) \times (W_{\mathcal{B},0}/\mathbf{H}^1_{\mathcal{B}}) \to W_{\mathcal{B},0}/\mathbf{H}^1_{\mathcal{B}} \tag{6.31}$$

As $Aut^1(\mathcal{B})$ is a finite group we deduce that a geometric quotient Q by this action exists. Now we quotient Q by $\mathbf{G}_{\mathcal{B}}/\mathbf{G}^1_{\mathcal{B}} \cong S_2$ and get the geometric quotient of $W_{\mathcal{B},\bullet}$. □

Case B.3. The graph $\mathcal{B}$ has exactly two vertices of the first generation and no one of the second. This means that if f is a polynomial having behaviour $\mathcal{B}$ then the curve C that it defines is not tangent to the line at infinity at any point. Let p_1 and p_2 be the points at infinity of f; consider $d := I_{p_1}(C,L)$, $n-d = I_{p_2}(C,L)$. We define $W_{\mathcal{B},0}$ by imposing $p_1 = (0:1:0)$ and $p_2 = (1:0:0)$ and, if $f(x,y) = \sum a_{i,j}x^iy^j$,

$$a_{d,n-d} = 1 \quad a_{d,n-d-1} = 0 \quad a_{d-1,n-d} = 0$$

Consider the induced groupoid $W_{\mathcal{B},\bullet}$. Let

$$\mathbf{H}_{\mathcal{B}} := \{(\lambda,\mu) : \lambda^d\mu^{n-d} = 1\}$$

We can define $\mathbf{G}_{\mathcal{B}}$ and an action of this group on $W_{\mathcal{B},0}$ exactly like in the case before. In this case also, the groupoid $W_{\mathcal{B},\bullet}$ is isomorphic to the groupoid associated with the action.

One checks easily that if f is in $W_{\mathcal{B},0}$ then its Newton polygon is in the square whose diagonal is the segment joinig $(0,0)$ and $(d, n-d)$. This diagonal divides the square in an upper triangle and a lower triangle. It is easy to study the orbits of this group action. We get:

(1) The orbit of f is of dimension 0 if and only if its Newton polygon is contained in the diagonal.
(2) The orbit of f is non closed if and only if its Newton polygon is contained in one of the triangles.

If we remove from $W_{\mathcal{B},0}$ the polynomials whose Newton polygon is contained in one of the triangles then we know that the quotient is separated. On the other hand, as $W_{\mathcal{B},0}$ is affine and $\mathbf{H}_{\mathcal{B}}$ reductive, by Geometric Invariant Theory we know that there exist an affine categotical quotient Q_W of the whole $W_{\mathcal{B},0}$ by the action of $\mathbf{H}_{\mathcal{B}}$. The set Y of polynomials whose Newton polygon is contained in one of the triangles is an invariant closed subset, and hence its image Z is an closed subset of Q. Hence we deduce that

$$Q \setminus Z$$

is a quasiaffine geometric quotient of $W_{\mathcal{B},0} \setminus Y$ by the action of $\mathbf{H}_{\mathcal{B}}$.

Case B.4. The graph $\mathcal{B}$ has exactly one vertex of the first generation. If f is any polynomial with behaviour graph $\mathcal{B}$ then the curve C that it defines is tangent to the line at infinity at this point (otherwise $\mathcal{B}$ would be the behaviour graph of a set of parallel lines). Let E_1, E_2 and $\tilde{C}_1$, $\tilde{C}_2$ be the exceptional divisors and the strict transforms of C by the blowing ups at $p_1 = C \cap L$ and $E_1 \cap L$. If we assume that $\tilde{C}_2$ does not meet E_2 at $L \cap E_2$ then it is easy to construct a primitive automorphism preserving the behaviour of f, which is impossible as we are assuming that $\mathbf{CT}(\mathcal{B})$ is trivial. We conclude that $\tilde{C}_2$ meets $L \cap E_2$.

The last diccussion implies that there exist vertices $w, v_1, v_2 \in \mathcal{B}$ such that w is the root, $p(v_1) = w$, $p(v_2) = \{w, v_1\}$ and there is an arrow attached to the edge conecting w and v_2.

We will consider several types of graphs:

Type 1.1: There is a free vertex v_k with $p(v_k)$ free of the first floor, $p(p(v_k))$ fixed and $p(p(v_k)) \neq v_2$; there is $v_m \in \mathcal{B}$ free, such that $p(v_m)$ is free of the second floor. We deduce that there is a vertex v_l that is a free central vertex of the second floor.

Type 1.2 There exist vertices like v_k and v_l as above, but not like v_m.

Type 1.3 There exists a vertex like v_k as above, but not like v_l or v_m.

Type 2.1: There is no vertex with the properties of v_k in last type 1; there exist two free vertices v_k, v_l such that $p(v_k) \neq p(v_l)$, $p(v_k)$ and $p(v_l)$ are free and $p(p(v_k)) = p(p(v_l)) = v_2$; there is $v_m \in \mathcal{B}$ free such that $p(v_m)$ is free of the second floor.

Type 2.2: There exist v_k and v_l like in last type, but not v_m.

Type 3: There exists a unique v_k with $p(v_k)$ free and $p(p(v_k)) = v_2$; there are no free vertices different from v_1, v_k and $p(v_k)$.

Type 4: There exists a free vertex v such that $p(v)$ is free and $\mathcal{B}$ is not of one of the types above.

Type 5: There is no free vertex v such that $p(v)$ is free.

We choose v_k, v_l and v_m to be the maximal with their property respect to $\leq_t$. Let $\{w, v_{i_1}, ..., v_{i_s} = v_k\}$, $\{w, v_{j_1}, ..., v_{j_t} = v_l\}$ and $\{w, v_{h_1}, ..., v_{h_u} = v_m\}$ be the sets of vertices respectively greater or equal than v_k, v_l and v_m by $\leq_p$.

Given $\mathcal{B}$ of any of the types before, consider $(I, \alpha) \in V_{\mathcal{B},\leq_a}$, for any vertex $v_i \in \mathcal{B}$ denote by x_i (resp. E_i) the point of I associated with v_i (resp. the exceptional divisor of the blowing up at x_i).

DEFINITION 6.39. Define $Z_{\mathcal{B}} \subset V_{\mathcal{B},\leq_a}$ to be the set of sequences $\{x_1, ..., x_n\} \in V_{\mathcal{B},\leq_a}$ such that the following properties hold when the involved infinitely near points appear: $x_1 = (0 : 1 : 0)$ and the coordinates of x_k, x_l and x_m respect to the canonical charts of $E_{i_{s-1}}$ and $E_{j_{t-1}}$ and $E_{h_{u-1}}$ are all of them 0.

Define W_0^1 by the formula $W_0^1 := \tau_{\mathcal{B},\leq_a}^{-1}(Z_{\mathcal{B}})$.

Let $l := I_{L\cap E_1}(\tilde{C}_1, E_1)$, $l' := I_{L\cap E_2}(\tilde{C}_2, E_2)$, $m := \text{mult}_{p_1}(C)$ and $m' := \text{mult}_{E_1\cap L}(\tilde{C}_1)$, (notice that l, m, l' and m' are determined by $\mathcal{B}$). If $\mathcal{B}$ is of type 4 or 5 we define $(N, M) := (m+m'-2l', n-m-m'+l')$; if $\mathcal{B}$ is of any other type then $(N, M) := (m-l, n-m)$. Let $(f, I_f\alpha) \in W_0^1$, suppose that $f(x, y) = \sum a_{i,j}x^iy^j$. We define $W_{\mathcal{B},\bullet} \subset W_0^1$ imposing the following equations: for all types $a_{N,M} = 1$ and $a_{n,0} = 1$; for all types except 1.1 and 2.1 impose $a_{N,M-1} = 0$; for types 1.3, 3, 4 and 5 impose $a_{N+1,M-1} = 0$; for types 4 and 5 impose $a_{n-1,0} = 0$.

Consider $(f, I_f, \alpha) \in W_{\mathcal{B},0}$. By Lemma 6.17 we obtain: for all types except 1.1 and 2.1, given any translation of the form $\theta = (x, y + b)$ we have $I_f = I_{f\circ\theta^{-1}}$; for types 1.3, 3 and 4, given any automorphism of the form $H(x, y) = (x, y + ax)$ we have $I_f = I_{f\circ H^{-1}}$; for type 5, given any tranlation of the form $\theta = (x + a, y)$ we have $I_f = I_{f\circ\theta^{-1}}$.

Consider the finite group (whose operation is pairwise multiplication)

$$\mathbf{H}_{\mathcal{B}} := \{(\lambda, \mu) : \lambda^n = \lambda^N\mu^M = 1\}$$

Define the group $\mathbf{G}_{\mathcal{B}}$ as the direct product $\mathbf{H}_{\mathcal{B}} \times Aut(\mathcal{B})$. There is an obvious group action $\rho : \mathbf{G}_{\mathcal{B}} \times W_{\mathcal{B},0} \to W_{\mathcal{B},0}$ defined as follows: given $((\lambda, \mu), \beta) \in \mathbf{G}_{\mathcal{B}}$ and $(f, I_f, \alpha) \in W_{\mathcal{B},0}$, let $H(x, y) := (\lambda x, \mu y)$, define

$$\rho(((\lambda, \mu), \beta), (f, I_f, \alpha)) := (f\circ H^{-1}, I_{f\circ H^{-1}}, \beta^{-1}\circ\alpha)$$

The following lemma is easy

LEMMA 6.40. *The groupoid $W_{\mathcal{B},\bullet}$ is isomorphic to the groupoid associated with the action of the finite group $\mathbf{G}_{\mathcal{B}}$ on $W_{\mathcal{B},0}$. Hence there exist an affine geometric quotient for the groupoid $W_{\mathcal{B},\bullet}$.*

6.2.4. Case C. The graph $\mathcal{B}$ is equivalent to $\mathcal{B}'$, the behaviour graph of a set of (may be multiple) parallel lines. We show directly that the functor of families $\mathcal{M}_{\mathcal{B}}$ is representable.

Using the methods Section 5.2 on groupoids one can show that the functors of families for $\mathcal{B}$ and $\mathcal{B}'$ are isomorphic. Hence we are reduced to the case in which $\mathcal{B}$ is the behaviour graph of a set of (may be multiple) parallel lines. Moreover we can reduce to families for which the point at infinity of any element of the family is $(0 : 1 : 0)$. Clearly any element of such a family is a polynomial that only involves the variable x. Consider two polynomials $f, g \in \mathbb{C}[x]$, and an automorphism $\phi \in Aut(\mathbb{C}^2)$ such that $f = g\circ\phi$, then the automorphism should be of the form $\phi(x, y) = (ax + b, \phi_2(x, y))$. As $g\circ\phi(x, y) = g(ax + b)$ we see that f and g are equivalent if and only if they are R-equivalent by an automorphism of the affine line. Let $f \in \mathbb{C}[x]$ it is clear that $f \in X_{\mathcal{B}}$ if and only if f has degree n, equal to the degree of $\mathcal{B}$. We consider the slice $W_{\mathcal{B},0}$, that consists of the monic polynomials in one variable of degree n such that the coefficient of x^{n-1} vanishes. This set of polynomials forms an affine space. If we restrict the action

$$Aut(\mathbb{C}[x]) \times \mathbb{C}[x] \to \mathbb{C}[x]$$

to $W_{\mathcal{B},0}$ we obtain a groupoid whose only arrows are the identity morphisms. Hence $W_{\mathcal{B},0}$ represents $\mathcal{M}_{\mathcal{B}}$.

6.2.5. The existence of moduli spaces and a classifying algorithm. In this subsection we record the main results obtained along the whole section, and we add some remarks about the algorithmic nature of our constructions. The following theorem is the existence theorem of the moduli spaces:

THEOREM 6.41. *Let $\mathcal{B}$ be any behaviour graph.*

- *If $\mathcal{B}$ is equivalent to a behaviour graph $\mathcal{B}'$ of minimal degree and type A then there exists an affine geometric quotient $\mathbf{M}_{\mathcal{B}}$ of $X_{\mathcal{B},\bullet}$, i.e, we can coarsely represent the functor $\mathcal{M}_{\mathcal{B}}$ by an affine algebraic variety $\mathbf{M}_{\mathcal{B}}$. Furthermore the quotient can be seen alternatively as:*
  1. *The quotient of the affine algebraic variety $W_{\mathcal{B}',0}$ by the action of the group $\mathbf{G}_{\mathcal{B}'}$.*
  2. *The successive quotient of the affine algebraic variety $W'_{\mathcal{B}',0}$ by the finite groups $\mathbf{H}_{\mathcal{B}'}$ and $Aut_{rm}(\mathcal{B}')$ if $\mathcal{B}'$ is of type $A.1$ or $A.2$ or by the finite groups $\mathbf{H}_{\mathcal{B}'}$ and $Aut_{rm}(\mathcal{B}'^{m})$ if $\mathcal{B}'$ is of type $A.3$.*
- *If $\mathcal{B}$ is equivalent to a behaviour graph $\mathcal{B}'$ of minimal degree and type $B.1$, $B.2$ or $B.4$ then there exists an affine geometric quotient $\mathbf{M}_{\mathcal{B}}$ of $X_{\mathcal{B},\bullet}$, i.e, we can coarsely represent the functor $\mathcal{M}_{\mathcal{B}}$ by an affine algebraic variety $\mathbf{M}_{\mathcal{B}}$. Furthermore the quotient can be seen as*
  1. *The successive quotient of the affine variety $W_{\mathcal{B}',0}$ by the finite groups $\mathbf{H}_{\mathcal{B}'}$ and $Aut(\mathcal{B}')$, if $\mathcal{B}'$ is of type B.1*
  2. *The successive quotient of the affine variety $W'_{\mathcal{B}',0}$ by the finite groups $\mathbf{H}_{\mathcal{B}'}$, $Aut^1(\mathcal{B}')$ and S_2 (the symmetric group of two elements, if $\mathcal{B}'$ is of type B.2.*
  3. *The quotient of the affine variety $W_{\mathcal{B}',0}$ by the finite group $\mathbf{G}_{\mathcal{B}'}$, if $\mathcal{B}'$ is of type $B.4$.*
- *If $\mathcal{B}$ is equivalent to a behaviour graph $\mathcal{B}'$ of minimal degree and type $B.3$ then the functor $\mathcal{M}_{\mathcal{B}}$ is not coarsely represented by a separated algebraic variety. Nevertheless the set U formed by the polynomials in $X_{\mathcal{B},0}$ that are right equivalent to a polynomial that can be the first component of an element of $W_{\mathcal{B}',0} \backslash Y$ (see the analysis of subcase B.3 to recall the definition of Y) is an open subset saturated by R-equivalence. The groupoid obtained by restriction of $X_{\mathcal{B},\bullet}$ to U has a quasiaffine geometric quotient.*
- *If $\mathcal{B}$ is equivalent to a behaviour graph of a set of parallel lines then, we can represent the functor $\mathcal{M}_{\mathcal{B}}$ by an affine space.*

PROOF. We only have to prove something in the case in which $\mathcal{B}$ is equivalent to a behaviour graph $\mathcal{B}'$ of minimal degree and type $B.3$. The only thing to be proved is that U is open. Consider the composition

$$p : W_{\mathcal{B}',0} \times_{f,s_{\mathcal{B}'}} X_{\mathcal{B}',1} \xrightarrow{proj.} X_{\mathcal{B}',1} \xrightarrow{t_{\mathcal{B}'}} X_{\mathcal{B}',0}$$

where f is the map from the slice $W_{\mathcal{B}',0}$ to $X_{\mathcal{B}',0}$. As p is universally open, if pr_1 denotes the first projection of $W_{\mathcal{B}',0} \times_{f,s_{\mathcal{B}'}} X_{\mathcal{B}',1}$, then

$$V := p(pr_1^{-1}(W_{\mathcal{B}',0} \setminus Y))$$

is an open subset of $X_{\mathcal{B}',0}$. Consider $C_\bullet$, the comparison groupoid between $\mathcal{B}$ and $\mathcal{B}'$, let s_C and t_C be its source and target maps. Consider the composition

$$p' : X_{\mathcal{B}',0} \times_{\iota,s_C} Y_1 \xrightarrow{proj.} Y_1 \xrightarrow{t_{\mathcal{B}'}} Y_0$$

where ι denotes the inclusion of $W_{\mathcal{B}',0}$ in Y_0. As p' is universally open, if π_1 denotes the first projection of $X_{\mathcal{B}',0} \times_{\iota,s_C} Y_1$ then $p'(\pi_1^{-1}(V))$ is an open subset of Y_0. The subset U is the intersection of $X_{\mathcal{B},0}$ with $p'(\pi_1^{-1}(V))$. This shows that U is open. □

Now we are going to present an algorithm that allows to decide if to arbitrarily given polynomials are R-equivalent or not, and, in case they are, provide an automorphism that takes one to the other. This is not the first time that such an algorithm is obtained. In [**28**] W. Neumann and P. Wightwick obtained one based on splice diagrams.

ALGORITHM 6.42. Let f, g be two elements of $\mathbb{C}[x, y]$. To decide whether they are R-equivalent we proceed in the following steps:

Step 1: compute their behaviours graphs $\mathcal{B}$ and $\mathcal{B}'$. It is possible to decide in finite time whether they are equivalent or not: if $\mathcal{B}$ and $\mathcal{B}'$ have n and n' vertices respectively then any composable triple $\mathcal{B}, (S, p), (S_0, p_0)$ such that the (S_0, p_0)-composition of (S, p) and $\mathcal{B}$ is $\mathcal{B}'$ should be such that S has no more than $n + m$ elements, hence there are a finite number of composable triples to try. If $\mathcal{B}$ and $\mathcal{B}'$ are not equivalent then f and g are not right equivalent. If $\mathcal{B}$ and $\mathcal{B}'$ are equivalent go to next step.

Step 2: as $\mathcal{B}$ and $\mathcal{B}'$ are equivalent we deduce that there exist a behaviour graph $\mathcal{B}''$ of minimal degree that is equivalent to both $\mathcal{B}$ and $\mathcal{B}'$. Choose a composable triple $\mathcal{B}, (S, p), (S_0, p_0)$ and such that the (S_0, p_0)-composition of (S, p) and $\mathcal{B}$ is $\mathcal{B}''$. Choose an automorphism $\alpha : \mathcal{B}_f \to \mathcal{B}$. The set $\alpha^{-1}(S_0)$ is a subset of I_f. We complete $\alpha^{-1}(S_0)$ to a saturated set S_f of infinitely near points whose parental system is isomorphic to (S, p) by adding infinitely near points that not belong to I_f. The set S_f is the set of infititely near points corresponding to an automorphism ψ of G. Then the polynomial $f \circ \psi^{-1}$ has behaviour graph $\mathcal{B}''$. Notice that all the steps that we have made up to now are constructive if we suppose that we know how to solve polynomial equations in one variable. Analogously we can compute an automorphism ϕ such that $g \circ \phi^{-1}$ has behaviour graph $\mathcal{B}''$. Consequently we can assume that f and g have the same behaviour graph $\mathcal{B}$ and that it is of minimal degree.

Step 3 Consider an admissible ordering in $\mathcal{B}$, choose α (resp. β), an isomorphism from $\mathcal{B}_f$ (resp. $\mathcal{B}_g$) to $\mathcal{B}$ and consider the triples (f, I_f, α), (g, I_g, β). In the slicing procedure we have described a constructive manner to find a triples (f', I'_f, α') and (g', I'_g, β') that are equivalent to (f, I_f, α) and (g, I_g, β) and that belong to

- the slice $W'_{\mathcal{B},0}$, if $\mathcal{B}$ is of type A and $B.2$.
- the slice $W_{\mathcal{B},0}$, if $\mathcal{B}$ is of type $B.1$, $B.3$, $B.4$ or C.

Suppose that $\mathcal{B}$ is not of type B.3. Looking to each case separatedly it is easy to see that the orbit $Orb((f', I'_f, \alpha')$ of (f', I'_f, α') by the groupoid $W'_{\mathcal{B},\bullet}$ (or $W_{\mathcal{B},\bullet}$ in the corresponding cases) is finite and can be effectively computed. In an analogous way we compute $Orb(g', I'_g, \beta')$. We have that: *The polynomials f and g are R-equivalent if and only if $Orb(f', I'_f, \alpha') = Orb(g', I'_g, \beta')$.*

It remains to analyse the case in which $\mathcal{B}$ is of type B.3. Returning to the terminology of subcase B.3 we recall that the Newton polygons of f' and g' should be contained in the square whose diagonal is the segment joining $(0,0)$ and $(d, n-d)$. This diagonal divides the square in an upper triangle and a lower triangle. If the Newton polygon of f' is contained in the diagonal and the Newton polygon of g' not we deduce that f and g are not R-equivalent. If the Newton polygon of both f' and g' is contained in the diagonal then f and g are R-equivalent if and only if $f' = g'$. It remains to analyze the case in which the Newton polygons of both f' and g' are not contained in the diagonal.

Let's assume first that $d \neq n - d$. Let (i, j) be a point of the square that is outside the diagonal such that the coefficient of $x^i y^j$ in f' does not vanish. If the coefficient of $x^i y^j$ in g' vanishes then f is not R-equivalent to g. Assume that the coefficient of $x^i y^j$ in g' does not vanish. Let $S(f')$ (resp. $S(g')$) to be the set of polynomials of the form $f' \circ H$ (resp. $g' \circ H$) where $H(x, y) = (\lambda x, \mu y)$, for $(\lambda, \mu) \in \mathbf{H}_{\mathcal{B}}$, such that the coefficient of $x^i y^j$ in f' (resp. g') equals to 1. We have that f and g are R-equivalent if and only if $S(f')$ equals $S(g')$.

Now we assume that $d = n - d$. Let (i, j) be a point of the square that is outside the diagonal such that the coefficient of $x^i y^j$ in f' does not vanish. Consider the coefficients of $x^i y^j$ and $x^j y^i$ in g'. If both of them vanish then f is not R-equivalent to g. If any of them is non zero let $S(f')$ (resp. $S(g')$) to be the set of polynomials of the form $f' \circ H$ (resp. $g' \circ H$) or $f' \circ \tau \circ H$ (resp. $g' \circ \tau \circ H$) where $\tau(x, y) = (y, x)$ and $H(x, y) = (\lambda x, \mu y)$, for $(\lambda, \mu) \in \mathbf{H}_{\mathcal{B}}$, such that the coefficient of $x^i y^j$ in f' (resp. g') equals to 1. We have that f and g are R-equivalent if and only if $S(f')$ equals $S(g')$. $\square$.

We have not worried for efficiency in the algorithm before. Our only aim is to show that our classifying method is constructive.

6.2.6. Related open problems. In the paper where Oka studied moduli of polynomials with one place at infinity (see [**29**]), appart from constructing moduli spaces for polynomials with a fixed sequence of weight vectors at infinity (or, that is the same, polynomials such that the embedded resolution graph of the singularity at infinity of its central fibre is a fixed one), he proves that for each $g \geq 0$ there are finitely many sequences of weight vectors such that an smooth curve of genus g with one marked point can be embedded in the affine plane with minimal degree and in such a way that its corresponding sequence of weight vectors is one of the given ones. Given a behaviour graph $\mathcal{B}$ and affine curve C, we say that $\mathcal{B}$ is *realizable by* C if there is an embedding of C in the affine plane whose behaviour graph is $\mathcal{B}$. We say that $\mathcal{B}$ is (g, r)-*realizable* if there is an smooth affine curve of genus g with r places at infinity that realizes $\mathcal{B}$.

PROBLEM 6.43. Fixed integers $g \geq 0$ and $r > 0$, how many non equivalent behaviours at infinity are (g, r)-realizable?

Fixed an behaviour graph $\mathcal{B}$, is there any (g, r) such that $\mathcal{B}$ is (g, r)-*realizable.* In other words, is there any $f \in X_{\mathcal{B}}$ such that $f = 0$ defines an smooth ireducible curve?

For work related to the second question see [**11**] ,[**12**], [**13**].

related to the last problem, but more difficult, is to study in how many non equivalent ways can we embed in $\mathbb{C}^2$ a given affine curve. This have been intensively studied, specially in the case of curves with one place at infinity. The first important result was the famous Abhyankar-Moh theorem (see [**1**]). Later Abhyankar and Sing proved in [**3**] that if C has one place at infinity then it has only a finite number of non equivalent embeddings of C in the affine plane. For more recent work in the subject see [**2**], [**36**], [**37**].

Suppose that $C \subset \mathbb{C}^2$ is an irreducible smooth affine plane curve. Let $\overline{C}$ be the completion of C to $\mathbb{P}^2$, and let $\eta : X \to \overline{C}$ be the normalisation. Let L be the line at infinity of $\mathbb{P}^2$. Consider the divisor $D_\infty := \eta^* L$ and call it *divisor at infinity* of C. Consider the linear systems $\mathcal{L} := \eta^* \mathcal{O}_{\mathbb{P}^2}(1)$ (the hyperplane sections) and $\mathcal{L}(D_\infty)$, the complete linear system associated with D_∞; then $\mathcal{L} \subset \mathcal{L}(D_\infty)$. Moreover, the affine reference system (x, y) induces a basis $(\eta^* X, \eta^* Y, \eta^* Z)$ of $\mathcal{L}$. The cuadruple $(X, D_\infty, \mathcal{L}, (\eta^* X, \eta^* Y, \eta^* Z))$ determines the embedding $C \subset \mathbb{C}^2$ and satisfies that D_∞ is the divisor associated with $\eta^* Z$. Consequently to give an embedding of C into $\mathbb{C}^2$ is equivalent to give a cuadruple $(X, D_\infty, \mathcal{L}, (X, Y, Z))$ where D_∞ is a divisor such that $X \setminus C$ is the support of D_∞, $\mathcal{L}$ is a linear system contained in $\mathcal{L}(D_\infty)$ and (X, Y, Z) is a basis of $\mathcal{L}$ such that D_∞ is the divisor associated with Z.

Suppose now that that C has several points at infinity, not merely more than one branch, i.e, that $\overline{C} \cap L$ has more than one point. Let $\phi \in Aut(\mathbb{C}^2)$ be an automorphism of degree 1; consider the embedding $\phi(C) \subset \mathbb{C}^2$. As ϕ extends to a projective automorphism of $\mathbb{P}^2$, the first three componets of the cuadruple associated with C are the same than the first three componets of the cuadruple associated with $\phi(C)$. Suppose now that the degree of ϕ is strictly bigger than one. Let D_∞ (resp. D'_∞) denote the divisors at infinity of C (resp. $\phi(C)$). Let $\{p_1, ..., p_r\}$ be the suppport of D_∞ and D'_∞; then $D_\infty = \sum_{i=1}^r n_i p_i$ and $D'_\infty = \sum_{i=1}^r m_i p_i$, where $n_i = I_{p_i}(C_i, L)$ and $m_i = I_{p_i}(C'_i, L)$, being C_i (resp. C'_i) the branch at infinity of C (resp. $\phi(C)$) centered at p_i. Using the techniques of section 4.1 it is easy to show that the sequences $n_1, ..., n_r$ and $m_1, ..., m_r$ cannot be equal, and hence that $D_\infty \neq D'_\infty$. Summarizing, we have seen that if $\phi \in Aut(\mathbb{C}^2)$ then either the divisors at infinity of C and $\phi(C)$ are different or the first three componets of the associated cuadruples are equal.

Let $C \subset \mathbb{C}^2$ be an smooth irreducible plane affine curve and $(X, D_\infty, \mathcal{L}, (X : Y : Z))$ its associated cuadruple. Consider $Grass(3, \mathcal{L}(D_\infty))$, the grassmanian of 3-dimensional subspaces in $\mathcal{L}(D_\infty)$; define $E(C)$ to be the locus of $Grass(3, \mathcal{L}(D_\infty))$ formed by linear systems $\mathcal{L}'$ that contain D_∞ and that if (X, Y, Z) is a basis of $\mathcal{L}'$ such that D_∞ is the divisor associated with Z, then $(X/Z, Y/Z) : C \to \mathbb{C}^2$ defines an embedding. Clearly $E(C)$ is not empty ($\mathcal{L} \in E(C)$), and, by the paragraph above, two points of $E(C)$ induce inequivalent embeddings of C. These arguments can be extended to a wide class of curves with one point and several branches at infinity.

The last discussion shows the relation of the methods of this thesis with

PROBLEM 6.44. Given any affine curve C, how many non equivalent behaviour (respectively embedded resolution) graphs can we obtain embedding $\mathcal{B}$ in different ways in the affine plane? In how many non equivalent ways can we embedd C? Are there infinitely many embeddings of C with the same behaviour graph?

6.3. On the geometry of the moduli spaces

Let $\mathcal{B}$ be a behaviour graph. In most of the cases the moduli space $\mathbf{M}_{\mathcal{B}}$ has been constructed as the quotient of an affine algebraic variety, that has been denoted by $W_{\mathcal{B},0}$ or $W'_{\mathcal{B},0}$ (depending of the type of graph), by successive actions of finite groups. Moreover we know how to compute the groups using only the combinatorics of $\mathcal{B}$. Given this it becomes clear that to investigate the geometry of $\mathbf{M}_{\mathcal{B}}$ it is useful to understand the geometry of $W_{\mathcal{B},0}$ or $W'_{\mathcal{B},0}$. In this section we will show how to give a quite close geometric description of these varieties using only the combinatorial information of $\mathcal{B}$.

NOTATION 6.45. To make the notation of this section more homogeneous we will write $W_{\mathcal{B},0}$ instead of $W'_{\mathcal{B},0}$.

A common feature of $W_{\mathcal{B},0}$ for all the cases considered is that for its construction we consider a closed subset $Z_{\mathcal{B}} \subset V_{\mathcal{B},\leq_a}$ obtained by fixing the position of the infinitely near points corresponding to a number of free vertices, and then we define a previous slice as the inverse image of $Z_{\mathcal{B}}$ by $\tau_{\mathcal{B},\leq_a}$. In this section we will denote $\tau^{-1}_{\mathcal{B},\leq_a}(Z_{\mathcal{B}})$ by W'_0. We define

$$\delta'_1(\mathcal{B}) := \sharp\{\text{infinitely near points fixed to construct} \quad Z_{\mathcal{B}}\} \tag{6.32}$$

The restriction to $Z_{\mathcal{B}}$ of the sequence of fibrations $V_{\delta_1(\mathcal{B})} \to \dots \to V_1$ introduced in Theorem 3.54 is such that the morphisms that correspond to deletion of vertices whose associated infinitely near point is fixed for the construction of $Z_{\mathcal{B}}$ become isomorphisms. Moreover as the first free vertex is one of them we deduce that the image of $Z_{\mathcal{B}}$ by the composition of all the morphisms of the sequence is just $(0:1:0) \in V_1$ (V_1 is the line at infinity). Thus, we get a sequence of locally trivial fibrations

$$V_{\delta_1(\mathcal{B})-\delta'_1(\mathcal{B})} \to \dots \to V_2 \to \{(0:1:0)\} \tag{6.33}$$

such that if $\nu_1, \dots, \nu_{\delta_1(\mathcal{B})-\delta'_1(\mathcal{B})}$ are the free vertices that do not correspond to infinitely near points that should be fixed, then each fibre of the i-th map is $\mathbb{C} \setminus \{\xi(\nu_i) \quad \text{points}\}$ (see Definition 3.53 to recall the value of $\xi(\nu_i)$).

Notice that for any behaviour $\mathcal{B}$ there exists sets $C_{\mathcal{B},1}$ and $C_{\mathcal{B},2}$ formed by pairs of integers such that $W_{\mathcal{B},0}$ is constructed in the following way: let $(f, I_f, \alpha) \in W'_0$, if $f(x,y) = \sum_{i,j} a_{i,j} x^i y^j$ then $(f, I_f, \alpha) \in W_{\mathcal{B},0}$ if and only if the following sets of conditions are satisfyied:

$$a_{i,j} = 0 \quad \text{if} \quad (i,j) \in C_{\mathcal{B},1} \tag{6.34}$$

$$a_{i,j} = 1 \quad \text{if} \quad (i,j) \in C_{\mathcal{B},2} \tag{6.35}$$

PROPOSITION 6.46. *Let $\mathcal{B}$ be a behaviour graph of minimal degree d, let $\leq_a$ be an admissible ordering on it. Consider any $(I,\alpha) \in V_{\mathcal{B},\leq_a}$. If $(i,j) \in C_{\mathcal{B},2}$ then the restriction of the equation $a_{i,j} = 0$ to the vector space $H^0(\mathbb{P}^2, \mathcal{I}_I(d))$ is a hyperplane that belongs to the arrangement that we have to delete from $H^0(\mathbb{P}^2, \mathcal{I}_I(d))$ to get $\tau^{-1}_{\mathcal{B},\leq_a}(I,\alpha)$. Hence the restriction of the equation $a_{i,j} = 1$ to $H^0(\mathbb{P}^2, I_I(d))$ is an affine hyperplane parallel to one of the hyperplanes of the arrangement.*

PROOF. The proof is a direct check that can be done case by case. □

PROPOSITION 6.47. *Let $\mathcal{B}$, $\leq_a$ and (I,α) like in Proposition 6.46. Assume that $\mathcal{B}$ is neither of case A.3, type 3 nor of case B.4, type 4. For any $(i,j) \in C_{\mathcal{B},1}$ denote by $H_{i,j}$ the hyperplane of $H^0(\mathbb{P}^2, \mathcal{O}_{\mathbb{P}^2}(d))$ defined by the restriction of the equation $a_{i,j} = 0$. Then the hyperplanes of the set*

$$\{H_v : v \in V_{min}(\mathcal{B})\} \cup \{H_{i,j} \cap H^0(\mathbb{P}^2, \mathcal{I}_I(d)) : (i,j) \in C_{\mathcal{B},1}\}$$

are linearly independent.

PROOF. Case A.3. is the most difficult one, as it involves all the reasonigs that are necessary for the other cases, we discuss it and leave the others to the reader.

Subcase A.3 We will use the notations of Proposition 3.48 and Lemma 3.51. We know the following assertions about $\mathcal{B}$:

(1) The parental system contains $w, v_1, ..., v_{r+1}$, a set of vertices for which the restriction of the parental function is the same that the restriction of the parental function of $\mathcal{A}_{r+1}$ to its first $r+2$ vertices.
(2) Any vertex $v \in \mathcal{B}$ different from $w, v_1, ..., v_{r+1}$ satisfy $v >_p v_{r+1}$.
(3) There are no arrows attached to the vertices $v_1, ..., v_r$ or to the edges connecting them. Moreover there exists an arrow a attached to the edge that connects v_r and v_{r+1}.

Give a total order $(i_1, j_1), ..., (i_t, j_t)$ in the set $C_{\mathcal{B},1}$. Consider the filtration

$$H^0(\mathbb{P}^2, \mathcal{O}_{\mathbb{P}^2}(d)) =: F''_{t+1} \supset F''_t \supset ... \supset F''_1$$

defined by $F''_h := (\cap_{l=h}^t) \cap H_{i_l,j_l}$. Define $F'_i := F''_i \cap H^0(\mathbb{P}^2, \mathcal{I}_I(d))$. Consider the filtration $F_k \supset ... \supset F_1$ introduced in Proposition 3.48. We will fix our attention in the case of $\mathcal{B}$ of type 4, which is the most difficult one. It is enough to show that

$$F'_{t+1} \supsetneq ... \supsetneq F'_1 \supsetneq F'_1 \cap F_k \supsetneq ... \supsetneq F'_1 \cap F_1$$

Consider the graph $\mathcal{B}_r$ obtained by deleting from $\mathcal{B}$ all the vertices except $w, v_1, ..., v_r$. It has only one arrow b attached to the edge connecting v_r with v_1, that comes from the deletion of the vertex v_{r+1}. Choose any polynomial $f \in \tau^{-1}_{\mathcal{B},\leq_a}(I,\alpha)$. Consider the blowing up process whose parental system is formed by $w, v_1, ..., v_r, v_{r+1}$ and whose first blowing up center is $p = (0:1:0)$. The weight of the arrow b of $\mathcal{B}_r$ determines the numbers $n = \deg(f)$, $m := \text{mult}_p(f)$, $l := I_{L \cap E_1}(\tilde{C}_1, E_1)$, $m' := \text{mult}_{E_1 \cap E_r}(\tilde{C}_r)$ and $l' := I_{E_r \cap E_{r+1}}(\tilde{C}_{r+1}, E_{r+1})$. Given these numbers, if $f(x,y) = \sum_{i+j \leq n} a_{i,j} x^i y^j$, we have that the conditions 6.23 are satisfied. Recall that we have defined $(N, M) := (rm - (r-1)n + rm' - (r+1)l', n - m - m' + l')$. With this terminology the set $C_{\mathcal{B},1}$ is equal to

$$\{(N+r, M-1), (N+r-1, M-1), ..., (N, M-1), (n-1, 0)\}$$

and we can consider it with the order that we have just given. Note that then $t = r + 2$.

Because of Lemma 6.32 and because of the asertions $1-3$ about $\mathcal{B}$ we know that for any $s \leq r$ and any $\phi \in Aut(\mathbb{C}^2)$ of the form

$$\phi(x,y) = (x + a, y + \sum_{i=0}^{s} b_i x^i) \tag{6.36}$$

we have that $I_{f \circ \phi^{-1}} = I_f = I$. As we know from the analysis of subcase A.3 that $f \circ \phi^{-1}$ have the same behaviour graph that f we conclude that $f \circ \phi^{-1} \in \tau^{-1}_{\mathcal{B},\leq_a}(I,\alpha)$. A computation that takes into account conditions 6.23 shows that for any $1 \leq u \leq t+1$ there exists an automorphism ϕ_u of the form 6.36 such that $f \circ \phi_u^{-1} \in F'_u \setminus F'_{u-1}$. As $f \circ \phi_1 \in \tau^{-1}_{\mathcal{B},\leq a}(I, \alpha)$ we deduce that it cannot be in F_k. Hence $f \circ \phi_1^{-1} \in F'_1 \setminus F'_1 \cap F_k$.

Because of Lemma 3.51 for any j such that $2 \leq j \leq k$ there exists an extended behaviour graph $\mathcal{G}_j$ and an element (I_j, α_j) of $V_{\mathcal{G}_j,\leq a}$ such that if $f \in \tau^{-1}_{\mathcal{G}_j,\leq a}(I_j, \alpha_j)$ then $f \in F_j \setminus F_{j-1}$. We will use the terminology Lemma 3.51 and consider several cases:

Case 1. The vertex v_{i_j} is neither v_{r+1} nor w nor v_1. As the vertices $v_1, ...v_r$ are in the dependency class of the root we deduce that v_{i_j} is strictly smaller than v_1 and v_{r+1} in the parental ordering. Consider the graph $\mathcal{H}$ obtained by deleting from $\mathcal{G}_l$ all the vertices except $w, v_1, ..., v_r$. It has only one arrow b' attached to the edge connecting v_r with v_1, that comes from the deletion of the vertex v_{r+1}. Condition 3 of Lemma 3.51 implies that $\mathcal{H}$ equals $\mathcal{B}_r$, the result of deleting all vertices of $\mathcal{B}$ except $w, v_1, ..., v_r$.

Let $f \in \tau^{-1}_{\mathcal{G}_l,\leq a}(I_j, \alpha_j)$. Consider the blowing up process whose parental system is formed by $w, v_1, ..., v_r, v_{r+1}$ and whose first blowing up center is $p = (0:1:0)$. The weight of the arrow b' of $\mathcal{H}$ determines the numbers $\deg(f)$, $\mathrm{mult}_p(f)$, $I_{L\cap E_1}(\tilde{C}_1, E_1)$, $\mathrm{mult}_{E_1 \cap E_r}(\tilde{C}_r)$ and $I_{E_r \cap E_{r+1}}(\tilde{C}_{r+1}, E_{r+1})$. The equality of $\mathcal{H}$ and $\mathcal{B}_r$ implies that these numbers are equal respectively to n, m, l, m' and l'. Hence, if $f(x,y) = \sum_{i+j \leq n} a_{i,j} x^i y^j$, then the conditions 6.23 are satisfied.

After this, an easy computation shows the existence of a $\phi \in Aut(\mathbb{C}^2)$ of the form 6.36 such that $f \circ \phi^{-1}$ is in F''_1. By the construction of I_j in Lemma 3.51 we have that $I \subset I_j = I_f$. Althoug it might be that $I_f \neq I_{f \circ \phi^{-1}}$, Lemma 6.32 shows that the points of $I \subset I_j$ are left fixed by ϕ. In the construction of I_j we had freedom in the choice of the points that we add to I to get I_j provided that the resulting parental function were of certain form. Using this freedom we can redefine $I_j := I_{f \circ \phi^{-1}}$. Then $f \circ \phi^{-1}$ is in $\tau^{-1}_{\mathcal{G}_j,\leq a}(I_j, \alpha_j)$ and hence f belongs to $F_j \setminus F_{j-1}$. Consequently $f \circ \phi^{-1} \in F'_1 \cap F_l \setminus F'_{l-1}$, as we needed.

Case 2. The vertex v_{i_j} equals v_{r+1}. Then v_{i_j+1} is not in the same dependency class than v_1. Consider the arrow a of $\mathcal{G}_{j+1}$ such that $\gamma(a) = \{v_r, v_{r+1}\}$; let (e, l') be its weight. Because of the case before any polynomial in $\tau^{-1}_{\mathcal{G}_{j+1},\leq a}(I_{j+1}, \alpha_{j+1})$ should satisfy the conditions 6.23. If $e \neq 1$ then, by an easy argument using Newton polygons, we deduce that the modifications introduced in $\mathcal{G}_{j+1}$ and (I_{l+1}, α_{j+1}) to construct $\mathcal{G}_l$ and (I_j, α_j) imply that if $f \in \tau^{-1}_{\mathcal{G}_j,\leq a}(I_j, \alpha_j)$ then the following conditions are satisfied:

(6.37)

$$\begin{array}{lll}
a_{i,j} = 0 & \text{if} & j > n - m \\
a_{i,j} = 0 & \text{if} & i + rj > n \\
a_{i,j} = 0 & \text{if} & i + rj = n \quad \text{and} \quad i \neq n \\
a_{n,0} \neq 0 & & \\
a_{i,j} = 0 & \text{if} & i + (r+1)j > 2n - m - m' - 1 \\
a_{i,j} = 0 & \text{if} & i + (r+1)j = 2n - m - m' - 1 \\
 & & \quad \text{and} \quad j < n - m - m' + l'
\end{array}$$

$$a_{rm-(r-1)n+rm'-(r+1)l'-1, n-m-m'+l'} \neq 0$$
$$a_{m-l,n-m} \neq 0$$

Looking at the Newton polygon it is easy to show the existence of a $\phi \in Aut(\mathbb{C}^2)$ of the form 6.36 such that $f \circ \phi^{-1}$ is in F_1''. As before, may be redefining I_j, we also know that $f \in F_l \setminus F_{l-1}$. Hence we have that $f \circ \phi^{-1} \in F_1' \cap F_j \setminus F_{j-1}'$.

It remains to study the case $e = 1$. In this case given any $f \in \tau^{-1}_{\mathcal{G}_j,\leq_a}(I_j, \alpha_j)$ its coeficients satisfy only the first 5 conditions of 6.37. Looking at the Newton polygon of f we see that what we need appart from these conditions to show the existance of $\phi \in Aut(\mathbb{C}^2)$ of the form 6.36 such that $f \circ \phi^{-1}$ is in F_1'' is that

$$(6.38) \qquad \begin{array}{l} a_{t,t'} := a_{rm-(r-1)n+rm'-(r+1)l'-1, n-m-m'+l} \neq 0 \\ a_{u,u'} :=_{rm-(r-1)n+rm'-(r+1)l'+r, n-m-m'+l-1} = 0 \end{array}$$

To achieve this we choose apropiately the infinitely near points that we have to add to I_{j+1} to get I_j: note that $t+(r+1)t' = u+(r+1)u' = 2n-m-m'-1$ and that the coefficients $a_{i,j}$ when $i+(r+1)j = 2n-m-m'-1$ are determined the the positions and mutiplicities of the intersections of E_{r+1} and $\tilde{C}_{r+1}$. When $e = 1$ one of the modifications that we have to introduce in $\mathcal{G}_{j+1}$ is to add a sequence of vertices $\nu_1, ..., \nu_{s+1}$ with parental function $p(\nu_{k+1}) = \nu_k$ for $k > 1$ and $p(\nu_1) = v_{i+1}$, and to attach an arrow a'' to ν_{s+1} with weight $l' + 1$. This could be substituted by adding two sequences of vertices $\nu_1, ..., \nu_{s+1}$ and $\mu_1, ..., \mu_{s+1}$with the same properties and two arrows with weights f and 1 respectively. Then to define I_j we would have to add infinitely near points for each one of this vertices. In particular we would have to chose two infinitely near points x_1 and x_2 within $\dot{E}_{r+1} \setminus I_{j+1}$. By choosing x_1 and x_2 we choose two of the intersection points of E_{r+1} with $\tilde{C}_{r+1}$, the others are determined by I_{j+1}. We can choose x_1 and x_2 such that the conditions 6.38 are satisfied.

Case 4. The vertex v_{i_j} equals v_1. In this case j equals 2. An argument using the Newton polygon shows that the polynomial $f = x^n$ belongs to $F_1' \cap F_2$. Obviously $x^n \notin F_1$.

Case 5. The vertex v_{i_j} equals w. In this case j equals 1. The polynomial $f = 1$ belongs to $F_1' \cap F_1$. This case finishes the proof. □

We can state now the main results of this section:

DEFINITION 6.48. Define $\delta_3(\mathcal{B})$ and $\delta_4(\mathcal{B})$ to be the number of elements of $C_{\mathcal{B},1}$ and $C_{\mathcal{B},2}$ respectively.

THEOREM 6.49. *Let $\mathcal{B}$ be any behaviour graph and let d be its degree. The morphism $\tau_{\mathcal{B},\leq_a} : W_{\mathcal{B},0} \to Z_{\mathcal{B}}$ is surjective and smooth of relative dimension*

$$(d+1)(d+2)/2 - \delta_2(\mathcal{B}) - \delta_3(\mathcal{B}) - \delta_4(\mathcal{B})$$

The variety $Z_{\mathcal{B}}$ is smooth of dimension $\delta_1(\mathcal{B}) - \delta_1'(\mathcal{B})$. Moreover there is a sequence

$$(6.39) \qquad Z_{\mathcal{B}} := V_{\delta_1(\mathcal{B})-\delta_1'(\mathcal{B})} \to ... \to V_2 \to \{(0:1:0)\}$$

of fibrations such that and for any $2 \leq i \leq \delta_1(\mathcal{B}) - \delta_1'(\mathcal{B})$ each fibre of π_i is $\mathbb{C} \setminus \{\xi(\nu_i) \text{ points}\}$ (where $\xi(\nu_i)$ is defined in Definition 3.53 and $\nu_1, ..., \nu_{\delta_1(\mathcal{B})-\delta_1'(\mathcal{B})}$ are the free vertices that do not correspond to infinitely near points fixed in the construction of $Z_{\mathcal{B}}$).

The variety $W_{\mathcal{B},0}$ is smooth and rational of dimension

$$dim(W_{\mathcal{B},0}) = \delta_1(\mathcal{B}) - \delta_1'(\mathcal{B}) + (d+1)(d+2)/2 - \delta_2(\mathcal{B}) - \delta_3(\mathcal{B}) - \delta_4(\mathcal{B})$$

If $\mathcal{B}$ is neither of case A.3, type 3 nor of case B.4, type 4, the mapping $\tau_{\mathcal{B},\leq_a}$ is a locally trivial fibration whose fibres are isomorphic to

$$\mathbb{C}^{(d+1)(d+2)/2-\delta_2(\mathcal{B})-\delta_3(\mathcal{B})-\epsilon(\mathcal{B})} \times (\mathbb{C}^*)^{\epsilon(\mathcal{B})-\delta_4(\mathcal{B})}$$

PROOF. The proof is not difficult after the previous work made in this section. □

COROLLARY 6.50. *If $\mathcal{B}$ is not of type B.3 then $\mathbf{M}_\mathcal{B}$ is the quotient of $W_{\mathcal{B},0}$ by a finite equivalence relation; we deduce that*

$$dim(\mathbf{M}_\mathcal{B}) = dim(W_{\mathcal{B},0}) = \delta_1(\mathcal{B}) - \delta_1'(\mathcal{B}) + (d+1)(d+2)/2 - \delta_2(\mathcal{B}) - \delta_3(\mathcal{B}) - \delta_4(\mathcal{B})$$

If $\mathcal{B}$ is of type B.3 then $dim(\mathbf{M}_\mathcal{B}) = dim(W_{\mathcal{B},0}) - 1$.

Consider the sequence: $W_{\mathcal{B},0} \overset{\tau_{\mathcal{B},\leq_a}}{\longrightarrow} V_{\delta_1(\mathcal{B})-\delta_1'(\mathcal{B})} \to \dots \to V_2 \to \{(0:1:0)\}$ If $\mathcal{B}$ is neither of case A.3, type 3 nor of case B.4, type 4, all the morphisms that appear are, from the topological point of view, locally trivial fibrations whose fibres are $K(\pi,1)$ spaces. Using iteratedly the long homotopy sequence of a fibration we deduce that

COROLLARY 6.51. *Let $\mathcal{B}$ be any behaviour graph neither of case A.3, type 3 nor of case B.4, type 4. If $i \geq 2$ then the i-th homotopy group of $W_{\mathcal{B},0}$ and $Z_\mathcal{B}$ vanishes.*

The last corollary implies the following beautiful topological property of families of polynomials:

COROLLARY 6.52. *Let $\mathcal{B}$ be any behaviour graph neither of case A.3, type 3 nor of case B.4 type 4; suppose that we have a continous family of ordered polynomials of $W_{\mathcal{B},0}$ parametrized over the sphere S^k, with $k \geq 2$. Then it can be extended to the ball that fills the sphere.*

An important question is to ask whether for a behaviour graph $\mathcal{B}$ the higher homotopy groups of $\mathbf{M}_\mathcal{B}$ vanish. When $\mathbf{M}_\mathcal{B}$ is the quotient of $W_{\mathcal{B},0}$ by succesive finite group actions, to prove the vanishing of the higher homotopy groups is enough to show that the actions are free. It turns out that, for most of the clases of graphs that we consider for the cutting of the slice there is an example of a graph $\mathcal{B}$ for which the actions are non free. Nevertheless the examples that we have found are very symmetric and have a low number of vertices. It seems that when the graphs are big enough the group actions are free.

PROBLEM 6.53. For which graphs $\mathcal{B}$ the higher homotopy groups of $\mathbf{M}_\mathcal{B}$ vanish?

APPENDIX A

Canonical orders

The aim of this appendix is to prove Proposition 2.55. Before we prove it in general we will deal with a particular type of graphs that will turn out to be the building blocks of any $\mathcal{A}$-constructible graph.

DEFINITION A.1. An *elementary parental system* of length k is a parental system (S,p) where

$$S = \{v_0, ..., v_k\} \bigcup \left(\bigcup_{i=0}^{k} \{w_1^i, ..., w_{r_i}^i\} \right)$$

for some natural numbers $r_1, ..., r_k$, and such that

$$p(v_1) = v_0$$

$$p(v_2) = \{v_0, v_1\}$$

$$p(v_i) = \{v_1, v_{i-1}\} \quad \text{for} \quad i \geq 3$$

$$p(w_j^i) = v_i \text{ for } 1 \leq j \leq r_i.$$

The vertices w_j^i are called *special* vertices of the elementary parental system and the v_i's are called *ordinary* vertices. An *elementary graph* is the graph associated with an elementary parental system.

CONSTRUCTION A.2. (Canonical orders in elementary graphs) Given any elementary parental system the parental ordering consists of the following relations: $v_i \leq_p v_j$ if and only if $i \geq j$ and $w_j^i \leq_p v_i$ for $1 \leq i \leq k$ and $1 \leq j \leq r_i$. We extend $\leq_p$ to an ordering $\leq$ by adding the following relations: $w_j^i \leq w_m^l$ if and only if $i \geq l$ and $v_i > w_k^j$ for all i, j, k.

Now let $(\mathcal{G}, \Lambda)$ be a decorated elementary graph. We use the decoration to extend the ordering among the vertices of the graph as follows: for fixed i consider two vertices w_j^i and w_l^i, let e_j (resp. e_l) the edge connecting w_j^i (resp. w_j^i) with v_i. The following procedure allows to compare w_j^i and w_l^i:

Step 1. If there is no arrow of Λ_4 attached to e_j and there is an arrow of Λ_4 attached to e_l then $w_j^i < w_l^i$. If there is no arrow of Λ_4 attached to either e_j or e_k then go to step 2. If there are arrows $a_j, a_l \in \Lambda_4$ attached to e_j, e_l respectively let ω_j, ω_l be their respective weights. If $\omega_j <_{lex} \omega_l$ then $w_j^i < w_l^i$. If $\omega_j = \omega_l$ then go to step 3.

Step 2. If there is no arrow of attached to e_j and there is an arrow of Λ_2 attached to e_l then $w_j^i < w_l^i$. If there is no arrow of Λ_2 attached to either e_j or e_k then go to step 3. If there are arrows $a_j, a_l \in \Lambda_2$ attached to e_j, e_l respectively let ω_j, ω_l denote their respective weights. If $\omega_j <_{lex} \omega_l$ then $w_j^i < w_l^i$. If $\omega_j = \omega_l$ then go to step 3.

Step 3. Let s_j, s_l be the number of arrows of Λ_3 attached to w^i_j, w^i_l respectively. Let $(\omega_{j,1}, ...\omega_{j,s_j})$, (resp. $(\omega_{l,1}, ...\omega_{l,s_l})$) be the weights of the arrows attached to w^i_j (resp. w^i_l) ordered lexicographically. If $s_j < s_l$ then $w^i_j < w^i_l$. Suppose $s_j = s_k$; if both numbers are zero go to step 4. In case they are not zero if $(\omega_{j,1}, ...\omega_{j,s_j}) <_{lex} (\omega_{l,1}, ...\omega_{l,s_l})$ then $w^i_j < w^i_l$, if $(\omega_{j,1}, ...\omega_{j,s_j}) = (\omega_{l,1}, ...\omega_{l,s_l})$ then w^i_j and w^i_l are not compared by the ordering.

Step 4. If there is no arrow of Λ_1 attached to w^i_j and there is an arrow of Λ_1 attached to w^i_l then $e_j < e_l$. If there is no arrow of Λ_1 attached to either w^i_j or w^i_k then w^i_j and w^i_l are not compared by the ordering. If there are arrows $a_j, a_l \in \Lambda_1$ attached to w^i_j, w^i_l respectively let ω_j, ω_l denote their respective weights. If $\omega_j < \omega_l$ then $w^i_j < w^i_l$. If $\omega_j = \omega_l$ then w^i_j and w^i_l are not compared by the ordering.

With this procedure we have defined a partial ordering in the set of vertices of $(\mathcal{G}, \Lambda)$ with the property that two vertices are not related if and only if they are $\sim_c$-equivalent. This gives automatically a partial ordering in the set of edges with the same property. Now we define an ordering in the set of arrows: we start by imposing that all the arrows attached to edges are bigger than all the arrows attached to vertices. Consider two arrows a, a' such that either both are attached to vertices or both are attached to edges, if $\gamma(a) < \gamma(a')$ then we define $a < a'$, if $\gamma(a)$ and $\gamma(a')$ are not related by the ordering then consider ω and ω', the weights of a and a'. If $\omega <_{lex} \omega'$ then $a < a'$. If $\omega = \omega'$ then a and a' are not related by the ordering. We have defined an ordering in the set of arrows such that two arrows are not related if and only if they are $\sim_c$ equivalent.

DEFINITION A.3. Let $(\mathcal{G}, \Lambda)$ be a decorated elementary graph. Define the canonical ordering on $(\mathcal{G}, \Lambda)$ to be the ordering constructed in the discussion above and denote it by $\leq_c$.

REMARK A.4. Let $(\mathcal{G}, \Lambda)$ be a decorated elementary graph. The following two facts are obvious from the construction of $\leq_c$.

(1) Two vertices or arrows of $(\mathcal{G}, \Lambda)$ are not related by the canonical ordering if and only if they are $\sim_c$-equivalent. Hence the canonical ordering defines a total ordering among the set of $\sim_c$-classes of a decorated elementary graph.
(2) If v, v' are vertices of $(\mathcal{G})$ and $v <_c v'$ for the canonical order associated with the graph without decoration $\mathcal{G}$ then $v <_c v'$ for the canonical order associated with the graph with decoration $(\mathcal{G}, \Lambda)$.

DISCUSSION A.5. (Total orders in elementary graphs) Consider a decorated elementary graph $(\mathcal{G}, \Lambda)$. Let $\leq_a$ be an admissible ordering on it. As the canonical ordering defined a total ordering of the $\sim_c$ classes of arrows and vertices, and the admissible ordering defines a total ordering of each equivalence class, we can define a total ordering $\leq_t$ in the sets of vertices and arrows as follows: let x, y be either vertices or arrows. We say that $x \leq_t y$ if and only if either $x <_c y$ or $x \sim_c y$ and $x \leq_a y$. Notice that ordering $\leq_t$ depends only on the decorated elementary graph and the admissible ordering. The restriction of $\leq_t$ to the graph without decoration $\mathcal{G}$ gives a total order on the set of vertices. From the second item of the last remark one deduces that there exists an admissible ordering $\leq'_a$ on the graph $\mathcal{G}$ such that the restriction from $\leq_t$ to $\mathcal{G}$ is the total ordering associated with $\leq'_a$ and the canonical order on $\mathcal{G}$.

Definition A.6. Let $(\mathcal{G}, \Lambda)$ be a decorated elementary graph and $\leq_a$ an admissible ordering on it. We say that $\leq_t$ is the total ordering associated with $\leq_a$.

Construction A.7. (An ordering in the set of elementary graphs) Consider 2 elementary graphs $(\mathcal{G}, \Lambda)$ and $(\mathcal{G}', \Lambda')$ of length k and k'. Use the notations of definition A.1. We compare them as follows:

Step 1. If $k < k'$ then $(\mathcal{G}, \Lambda) < (\mathcal{G}', \Lambda')$. Assume that $k = k'$. If $(r_1, ..., r_k) <_{lex} (r'_1, ..., r'_k)$ then $(\mathcal{G}, \Lambda) < (\mathcal{G}', \Lambda')$. If $(r_1, ..., r_k) = (r'_1, ..., r'_k)$ then $\mathcal{G}$ is isomorphic to $\mathcal{G}'$. Go to step 2.

Step 2. First we describe a method to compare graphs consisting of a decorated vertex: consider 2 decorated graphs with only one vertex. We denote them by v, v' (just the name of the unique vertex, omitting the names of the arrows). We compare them as follows: if neither v nor v' have arrows then we say that they are not related. If v does not have arrows and v' has an arrow of Λ_1 then $v < v'$. Let v, v' have each of them an arrow of Λ_1 with weights ω and ω' respectively. If $\omega < \omega'$ we say that $v < v'$. If $\omega = \omega'$ then we say that v and v' are not related. If v has less arrows of Λ_2 than v' then $v < v'$. Let assume that both v and v' have s arrows of Λ_2, let $(\omega_1, ..., \omega_s)$ (resp. $(\omega'_1, ..., \omega'_s)$) be the weights of the arrows of v (resp. v') lexicographically ordered. If $(\omega_1, ..., \omega_s) <_{lex} (\omega'_1, ..., \omega'_s)$ then we say that $v < v'$. If $(\omega_1, ..., \omega_s) = (\omega'_1, ..., \omega'_s)$ then we say that v and v' are not related.

We continue with the comparison of the graphs. Let $\mathcal{V}_i$ be the graph formed only by the vertex v_i and its arrows. Let $\mathcal{V}'_i$ be the graph formed only by the vertex v_i and its arrows. If $(\mathcal{V}_1, ..., \mathcal{V}_k) <_{lex} (\mathcal{V}'_1, ..., \mathcal{V}'_k)$ then we say that $(\mathcal{G}, \Lambda) < (\mathcal{G}', \Lambda')$. If $(\mathcal{V}_1, ..., \mathcal{V}_k) = (\mathcal{V}'_1, ..., \mathcal{V}'_k)$ go to step 3.

Step 3. First we describe a method to compare edges of the two different elementary graphs: given $0 \leq i < j \leq k$, we let $e_{i,j}$ (resp. $e'_{i,j}$) be the edge joining the vertexes v_i and v_j (resp. v'_i and v'_j) of $(\mathcal{G}, \Lambda)$ (resp. $(\mathcal{G}', \Lambda')$. Fixed i, j as above, if there is no arrow attached neither to $e_{i,j}$ nor to $e'_{i,j}$ we say that both edges are not related. If there is no arrow attached to $e_{i,j}$ and there is an arrow of Λ_3 attached to $e'_{i,j}$ we say that $e_{i,j} < e'_{i,j}$. If there is one arrow of Λ_3 (resp. Λ'_3) with weight ω (resp. ω') attached $e_{i,j}$ (resp. $e'_{i,j}$) and $\omega < \omega'$ we say that $e_{i,j} < e'_{i,j}$. If $\omega = \omega'$ we say that $e_{i,j}$ and $e'_{i,j}$ are not related. If there is no arrow of Λ_4 attached to $e_{i,j}$ and there is an arrow of Λ_4 attached to $e'_{i,j}$ we say that $e_{i,j} < e'_{i,j}$. If there is one arrow of Λ_4 (resp. Λ'_4) with weight ω (resp. ω') attached $e_{i,j}$ (resp. $e'_{i,j}$) and $\omega < \omega'$ we say that $e_{i,j} < e'_{i,j}$. If $\omega = \omega'$ we say that $e_{i,j}$ and $e'_{i,j}$ are not related.

Now we can continue with the comparison between the graphs. Assume that $e_{i,j} < e'_{i,j}$ and that for each $(l, m) <_{lex} (i, j)$ the edges $e_{l,m}$ and $e'_{l,m}$ are not comparable. Then $(\mathcal{G}, \Lambda) < (\mathcal{G}', \Lambda')$. If for any i, j such that $0 \leq i < j \leq k$ and that the edges $e_{i,j}$ and $e'_{i,j}$ exist it happens that they are not comparable go to step 4.

Step 4. Denote by e^i_j be the edge connecting w^i_j with v_i. As we have described a way to order edges and single vertices with decoration we can consider the lexicographically ordered sequence of pairs $((e^i_1, w^i_1), ..., (e^i_{r_i}, w^i_{r_i}))$ for each v_i of $(\mathcal{G}, \Lambda)$. We can do the same for each v'_i of $(\mathcal{G}', \Lambda')$. Assume that for any $j < i$ we have that

$$((e^j_1, w^j_1), ..., (e^j_{r_j}, w^j_{r_j})) = ((e'^j_1, w'^j_1), ..., (e'^j_{r_j}, w'^j_{r_j}))$$

and that

$$((e^i_1, w^i_1), ..., (e^i_{r_i}, w^i_{r_i})) <_{lex} ((e^i_1, w^i_1), ..., (e^i_{r_i}, w^i_{r_i}))$$

then we say that $(\mathcal{G},\Lambda) < (\mathcal{G}',\Lambda')$. If for any $i \leq k$ we have that

$$((e_1^i, w_1^i), ..., (e_{r_i}^i, w_{r_i}^i)) = ((e_1^i, w_1^i), ..., (e_{r_i}^i, w_{r_i}^i))$$

then we say that $(\mathcal{G},\Lambda)$ and $(\mathcal{G}',\Lambda')$ are not related. This finishes the comparison of the graphs.

NOTATION A.8. In the construction above we have defined a partial order in the set of elementary graphs. We will denote it by $\leq_g$.

REMARK A.9. Let $(\mathcal{G},\Lambda)$ and $(\mathcal{G}',\Lambda')$ be elementary graphs. The following facts are obvious from the way of constructing $\leq_g$:

(1) $(\mathcal{G},\Lambda)$ and $(\mathcal{G}',\Lambda')$ are not related by the relation $\leq_g$ if and only if they are isomorphic and not equal. Hence $\leq_g$ define a total ordering in the set of graphs modulo isomorphism.
(2) If $\mathcal{G} <_g \mathcal{G}'$ then $(\mathcal{G},\Lambda) <_g (\mathcal{G}',\Lambda')$.

Consider a decorated elementary graph $(\mathcal{G},\Lambda)$. Let $\leq_a$ be an admissible ordering on it and $\leq_t$ the total ordering obtained from the canonical order and $\leq_a$. Let $(\mathcal{G}_0,\Lambda_0)$ be the decorated elementary graph obtained from the deletion of the last (smallest) vertex. As we have a total ordering $\leq_t$ in $(\mathcal{G},\Lambda)$ it is clear that the vertices of $(\mathcal{G}_0,\Lambda_0)$ inherit a total ordering $\leq_t'$. If we examine the procedure for constructing the decoration of a graph after having deleted a vertex we see that all the arrows of Λ_0 except for possibly one were arrows of Λ. We give a total order $\leq_t'$ to the set Λ_0 as follows: impose that the restriction of $\leq_t'$ to the set of arrows of Λ_0 that were an arrow of Λ coincides with the restriction of $\leq_t$ to the same set. Notice that if $\leq_c'$ is the canonical order of $(\mathcal{G}_0,\Lambda_0)$, then, for any 2 arrows of this set such that $a <_c' a'$, it also holds that $a <_t' a'$. Let $a \in \Lambda'$ be the only arrow that was not an arrow of Λ. We complete $\leq_t'$ to a total order by imposing that a is the first arrow in its $\sim_c$-class. The following remarks follow very easily:

REMARK A.10. (*Orderings and deletion of the last vertex*)

(1) Let $(\mathcal{G},\Lambda)$ be a decorated elementary graph and $\leq_a$ an admissible ordering on it. Let $\leq_t$ be the total ordering resulting from the admissible ordering $\leq_a$ and the canonical ordering on $(\mathcal{G},\Lambda)$. If $(\mathcal{G}_0,\Lambda_0)$ and $\leq_t'$ are as in the discussion above then there exist an admissible ordering $\leq_a'$ on $(\mathcal{G}_0,\Lambda_0)$ such that $\leq_t'$ is the total order resulting from the canonical order $\leq_c'$ on $(\mathcal{G}_0,\Lambda_0)$ and the admissible order $\leq_a'$.
(2) $(\mathcal{G}_0,\Lambda_0) <_g (\mathcal{G},\Lambda)$

DISCUSSION A.11. Now we describe how any decorated $\mathcal{A}$-constructible graph $(\mathcal{G},\Lambda)$ can be constructed by successively adding decorated elementary graphs: let (S,p) parental system of $\mathcal{G}$; we see $(\mathcal{G},\Lambda)$ as a graph constructed in several steps as follows:

Step 0. Consider the subset S_0 formed by the root w_0 and all the vertices v such that $p(v) = w_0$. Let $(\mathcal{G}_0,\Lambda_0)$ be the graph obtained from $(\mathcal{G},\Lambda)$ by deletion of all the vertices of $S \setminus S_0$. Notice that $(\mathcal{G}_0,\Lambda_0)$ is an elementary graph of length 0. Let $S_0 = \{w_0, v_1, ..., v_r\}$, the vertices $\{v_1, ..., v_r\}$ are the *special vertices* of S_0.

Step k. Assume that we have defined the decorated graph $(\mathcal{G}_{k-1},\Lambda_{k-1})$ (if $k = 1$ and we have specified a subset $\{w_1, ..., w_s\}$ of vertices that are declared to be the special vertices of $\mathcal{G}_{k-1}$. For each special vertex w_i consider the set $S_{k,i}$ defined to be the unique maximal subset of S such that it contains w_i and $p(w_i)$,

p transforms $S_{k,i} \setminus \{p(w_i)\}$ into $S_{k,i}$ and the pair $(S_{k,i}, p)$ forms an elementary parental system such that $p(w_i)$ and w_i are its 0-th and first *ordinary* vertices,and that there are no special vertices w such that $p(w) = p(w_i)$. Define

$$S_k := S_{k-1} \bigcup \left(\cup_{i=1}^{r} S_{k,i} \right)$$

and let $(\mathcal{G}_k, \Lambda_k)$ be the graph obtained from deleting from $(\mathcal{G}, \Lambda)$ all the vertices not contained in S_k. The set of special vertices of S_k is defined to be the union of the set of special vertices of each of the $S_{k,i}$'s. It is clear that after a finite number steps $(\mathcal{G}, \Lambda) = (\mathcal{G}_n, \Lambda_n)$.

DEFINITION A.12. Recall from Definition 2.54 that $\mathbf{G}$ denotes the set of decorated $\mathcal{A}$-constructible graphs. We let $\mathbf{G_i}$ denote the subset of graphs of $\mathbf{G}$ constructible in a number of steps smaller or equal than i.

After these preliminaries we can finally prove the proposition.

PROOF OF PROPOSITION 2.55. Denote by $\mathbf{G^1_k}$ the set of graphs constructible in k steps such that, if w is the root there is a unique v such that $p(v) = w$. To prove the proposition it is enough to show, that, for any $k \in \mathbb{N}$, we can define a canonical order for each graph of $\mathbf{G_k} \bigcup \mathbf{G^1_{k+1}}$ and a partial ordering $\leq_g$ in $\mathbf{G_k} \bigcup \mathbf{G^1_{k+1}}$ such that the properties $1-7$ hold. We do it by induction on k:

Initial step. Any graph of $\mathbf{G_0}$ or $\mathbf{G^1_1}$ is an elementary parental graph. We have already defined a canonical order for each elementary graph and an ordering $\leq_g$ on the set of decorated elementary graphs that satisfy conditions $1-7$.

Inductive step. Assume that we have defined a canonical order for each graph of $\mathbf{G_k} \bigcup \mathbf{G^1_{k+1}}$ and a partial ordering $\leq_g$ in $\mathbf{G_k} \bigcup \mathbf{G^1_{k+1}}$ such that the properties $1-7$ hold. We have to define a canonical order for each graph of $\mathbf{G_{k+1}} \bigcup \mathbf{G^1_{k+2}}$ and a partial order on $\mathbf{G_{k+1}} \bigcup \mathbf{G^1_{k+2}}$ with the same properties. Consider $(\mathcal{G}, \Lambda) \in \mathbf{G_{k+1}} \bigcup \mathbf{G^1_{k+2}}$. If $(\mathcal{G}, \Lambda) \in \mathbf{G_k} \bigcup \mathbf{G^1_{k+1}}$ then we already have a canonical ordering in it satisfying the properties 1, 2, 4 and 6. If $(\mathcal{G}, \Lambda) \in \mathbf{G_{k+1}}$ we define $\leq_c$ on it as follows:

Let w be the root of $(\mathcal{G}, \Lambda)$, for any vertex v such that $p(v) = w$ we can define the graph $(\mathcal{H}_v, \Lambda_v)$ as the graph constructed by deleting first all the arrows not attached to a vertex or an edge that is descendant of v and then all the vertices that are not descendants of v except w (this operation consists in taking only the branch of the graph $(\mathcal{G}, \Lambda)$ that grows from v). We consider also the graphs $\mathcal{G}$ and $\mathcal{H}_v$'s obtained from the previous graphs by deletion of decorations. Notice that the graphs $(\mathcal{H}_v, \Lambda_v)$'s and $\mathcal{H}_v$'s are in $\mathbf{G^1_{k+1}}$, where , by induction, the order $\leq_g$ is already defined among them. Now we proceed in two steps:

Step 1. Let v_1 and v_2 be vertices such that $p(v_1) = p(v_2) = w$. Let v_1' and v_2' be vertices such that $v_i' \leq_p v_i$ for $i \in \{1, 2\}$. If $\mathcal{H}_{v_1} <_g \mathcal{H}_{v_2}$ then we impose $v_1' <_c v_2'$. If $\mathcal{H}_{v_1}$ and $\mathcal{H}_{v_2}$ are not related by the order $\leq_g$, then by induction it follows that they are isomorphic, consider any isomorphism α between them, we say that $v_1' <_c v_2'$ if $\alpha(v_1') <_c v_2'$ (the canonical order is defined in $\mathcal{H}_{v_2}$. If $\alpha(v_1')$ and v_2' are not related then go to step 2. We show that this definition is independent from the choice of α: by induction we know that if v, v', v'' are vertices of $\mathcal{H}_{w_2}$ such that $v \sim_c v'$ then $v <_c v''$ if and only if $v' <_c v''$. If α' is an isomorphism different from α then $\alpha'(v_1') = \alpha' \circ \alpha^{-1}(\alpha(v_1'))$, being $\alpha' \circ \alpha^{-1}$ an automorphism of $\mathcal{H}_{w_2}$. Hence $\alpha(v_1') \sim_c \alpha'(v_1')$ and then $\alpha(v_1') <_c v_2'$ if and only if $\alpha'(v_1') <_c v_2'$.

Step 2. If $(\mathcal{H}_{v_1}, \Lambda_{v_1}) <_g (\mathcal{H}_{v_2}, \Lambda_{v_2})$ then we impose $v_1' <_c v_2'$, if $(\mathcal{H}_{v_1}, \Lambda_{v_1})$ and $(\mathcal{H}_{v_2}, \Lambda_{v_2})$ are equal or not related by the order $\leq_g$, then by induction it follows that they are isomorphic, consider any isomorphism α between them, we say that $v_1' \leq_c v_2'$ if and only if $\alpha(v_1') \leq_c v_2'$ (the canonical order is defined in $(\mathcal{H}_{v_2}, \Lambda_{v_2})$ by induction). The fact that the definition is independent from the choice of α follows from the same reason used in step 1.

We define $\leq_c$ in the set of arrows following exactly the same procedure for the definition of the canonical order at the set of arrows of an elementary graph.

The ordering $\leq_c$ clearly satisfies the property 1. We claim that it also satisfies the property 2. We will prove it only for vertices and leave to the reader the case of arrows that is analogous. Let v_1', v_2' be two different vertices that are related by an automorphism, it is easy to show that v_1' and v_2' are not related by $\leq_c$ and we leave it to the reader. For the converse we consider two vertices v_1' and v_2' that are not related by $\leq_c$. Let v_1, v_2 be the vertices such that $v_1' \in \mathcal{H}_{v_1}$ and $v_2' \in \mathcal{H}_{v_2}$. It follows that $(\mathcal{H}_{v_1}, \Lambda_{v_1})$ and $(\mathcal{H}_{v_2}, \Lambda_{v_2})$ are not related by $\leq_g$ and hence they are isomorphic. Let $\alpha : (\mathcal{H}_{v_1}, \Lambda_{v_1}) \to (\mathcal{H}_{v_2}, \Lambda_{v_2})$ such an isomorphism, as v_1', v_2' are not related by $\leq_c$ we conclude that $\alpha(v_1')$ and v_2' are related by an automorphism of $(\mathcal{H}_{v_2'}, \Lambda_{v_2'})$. Composing α with that automorphism we can assume that $\alpha(v_1') = v_2'$. We can regard $(\mathcal{H}_{v_1}, \Lambda_{v_1})$ and $(\mathcal{H}_{v_2}, \Lambda_{v_2})$ as subsets of $(\mathcal{G}, \Lambda)$ and extend α to an automorphism $\beta \in Aut(\mathcal{G}, \Lambda)$ by letting fixed the root and all the vertices and arrows that are not either in $(\mathcal{H}_{v_1}, \Lambda_{v_1})$ or $(\mathcal{H}_{v_2}, \Lambda_{v_2})$, and imposing that the restriction of β to $(\mathcal{H}_{v_1}, \Lambda_{v_1})$ is α and the restriction of β to $(\mathcal{H}_{v_2}, \Lambda_{v_2})$ is α^{-1}. Clearly $\beta(v) = v'$ and hence $v \sim_c v'$.

We show now that the ordering $\leq_c$ satisfies property 4. Consider an admissible ordering $\leq_a$ on $(\mathcal{G}, \Lambda)$ and let $\leq_t$ be the total ordering obtained from $\leq_c$ and $\leq_a$. Let $(\mathcal{G}_0, \Lambda_0)$ be the graph obtained by the deletion of the smallest vertex and $\leq_t'$ the total order obtained by the restriction of $\leq_t$. To show that there exists an admissible order $\leq_a'$ on $(\mathcal{G}', \Lambda')$ such that $\leq_t$ is the order obtained from $\leq_a'$ and the canonical order $\leq_c'$ in $(\mathcal{G}_0, \Lambda_0)$ we only need to show that $\leq_t'$ extends $\leq_c'$. Let x be last vertex of $(\mathcal{G}, \Lambda)$. There is a unique v such that $x \in \mathcal{H}_v$. As x is the smallest vertex it follows that if v' is any vertex such that $p(v')$ is the root of $(\mathcal{G}, \Lambda)$ then either $(\mathcal{H}_v, \Lambda_v) <_g (\mathcal{H}_{v'}, \Lambda_{v'})$ or $(\mathcal{H}_v, \Lambda_v)$ and $(\mathcal{H}_{v'}, \Lambda_{v'})$ are isomorphic. By induction hypothesis, if we denote by $(\mathcal{H}_{v,0}, \Lambda_{v,0}')$ the result of deleting x from $(\mathcal{H}_v, \Lambda_v)$ we have that $(\mathcal{H}_{v,0}, \Lambda_{v,0}') <_g (\mathcal{H}_v, \Lambda_v)$, and hence $(\mathcal{H}_{v,0}, \Lambda_{v,0}') <_g (\mathcal{H}_{v'}, \Lambda_{v'})$ for any v'. This implies that any vertex of $(\mathcal{H}_{v,0}, \Lambda_{v,0}')$ is smaller by $\leq_c'$ than any vertex of any $(\mathcal{H}_{v'}, \Lambda_{v'})$ for $v' \neq v$. The same happens for $\leq_t'$. Hence to show that $\leq_t'$ extends $\leq_c'$ we can work separatedly on $(\mathcal{H}_{v,0}, \Lambda_{v,0})$ and on its complementary in $(\mathcal{G}_0, \Lambda_0)$. On $(\mathcal{H}_{v,0}, \Lambda_{v,0})$ the result follows by induction hypothesis and on its complementary both $\leq_c'$ and $\leq_t'$ are the restrictions of $\leq_c$ and $\leq_t$ respectively, and $\leq_t$ is an extension of $\leq_c$.

The fact that $\leq_c$ satisfies property 6 can be proved easily using that given a graph $(\mathcal{G}, \Lambda)$ its canonical order is constructed in two steps, and after the first the order that have been constructed is precisely the canonical order in the graph without decoration $\mathcal{G}$.

If $(\mathcal{G}, \Lambda) \in \mathbf{G}^{\mathbf{1}}_{\mathbf{k+2}}$ we can define $\leq_c$ with the properties 1, 2, 4 and 6 in an analogous way, but considering instead of the $(\mathcal{H}_v, \Lambda_v)$'s the graphs $(\mathcal{H}_{w_j^i}, \Lambda_{w_j^i})$, for w_j^i the special vertices of the elementary graph obtained by considering $(\mathcal{G}_1, \Lambda_1)$, the first step in the construction of $(\mathcal{G}, \Lambda)$.

We define now $\leq_g$ on $\mathbf{G_{k+1}} \bigcup \mathbf{G^1_{k+2}}$. We start by imposing that $\leq_g$ extends the ordering defined by induction on $\mathbf{G_k} \bigcup \mathbf{G^1_{k+1}}$ and that any graph of $\mathbf{G_{i+1}} \setminus \mathbf{G_i}$ is strictly bigger than any graph of $\mathbf{G_i}$ for any $i \geq 1$. Let $(\mathcal{G}, \Lambda)$ and $(\mathcal{G}', \Lambda')$ two graphs of $\mathbf{G_{k+1}} \setminus \mathbf{G_k}$, we compare them as follows: Let n, n' be the number of preimages by p of the root of $(\mathcal{G}, \Lambda)$, $(\mathcal{G}', \Lambda')$ respectively; if $n > n'$ then we say that $(\mathcal{G}, \Lambda) >_g (\mathcal{G}', \Lambda')$. Assume $n = n'$. We define $\leq_g$ in two steps:

Step 1. Delete the decorations. By induction all the $\mathcal{H}_v$'s of $\mathcal{G}$ can be compared by $\leq_g$, and we can consider the sequence $(\mathcal{H}_{v_1}, ..., \mathcal{H}_{v_n})$, ordered increasingly by $\leq_g$. We can consider $(\mathcal{H}'_{v'_1}, ..., \mathcal{H}'_{v'_n})$, the analogous sequence for $\mathcal{G}'$. If $(\mathcal{H}_{v_1}, ..., \mathcal{H}_{v_n}) <_{lex} (\mathcal{H}'_{v'_1}, ..., \mathcal{H}'_{v'_n})$ we say that $(\mathcal{G}, \Lambda) <_g (\mathcal{G}', \Lambda')$. Assume $\mathcal{H}_{v_i}$ is isomophic to $\mathcal{H}_{v'_i}$ for any $i \leq n$. Then it is obvious that the graphs without decoration $\mathcal{G}$ and $\mathcal{G}'$ are isomorphic. Go to step 2.

Step 2. This step runs parallel to step one but uses decorations: by induction all the $(\mathcal{H}_v, \Lambda_v)$'s of $(\mathcal{G}, \Lambda)$ can be compared by $\leq_g$, and we can consider the sequence $((\mathcal{H}_{v_1}, \Lambda_{v_1}), ..., (\mathcal{H}_{v_n} \Lambda_{v_n}))$, ordered increasingly by $\leq_g$. We can consider $((\mathcal{H}'_{v'_1}, \Lambda'_{v'_1}), ..., (\mathcal{H}'_{v_n} \Lambda'_{v'_n}))$, the analogous sequence for $(\mathcal{G}', \Lambda')$. If $((\mathcal{H}_{v_1}, \Lambda_{v_1}), ..., (\mathcal{H}_{v_n} \Lambda_{v_n})) <_{lex} ((\mathcal{H}'_{v'_1}, \Lambda'_{v'_1}), ..., (\mathcal{H}'_{v_n} \Lambda'_{v'_n}))$ we say that $(\mathcal{G}, \Lambda) <_g (\mathcal{G}', \Lambda')$. Assume the contrary: $(\mathcal{H}_{v_i}, \Lambda_{v_i})$ is isomorphic to $(\mathcal{H}'_{v'_i}, \Lambda'_{v'_i})$ for all $i \leq n$, then we consider the graphs $(\mathcal{G}_*, \Lambda_*)$, $(\mathcal{G}'_*, \Lambda'_*)$, the deletion of all the vertices minus the root for each of the two graphs. We say that $(\mathcal{G}, \Lambda) \leq_g (\mathcal{G}', \Lambda')$ if and only if $(\mathcal{G}_*, \Lambda_*) \leq_g ((\mathcal{G}'_*, \Lambda'_*)$.

The fact that $\leq_g$ satisfies properties 3 and 5 is easy. Property 7 follows from the fact that we have defined $\leq_g$ in two steps and the first is enough to define $\leq_g$ for graphs without decoration.

To define $\leq_g$ on $\bigcup \mathbf{G^1_{k+2}}$ so that the properties 3, 5 and 7 hold we proceed analogously as in the paragraph before but considering, instead of the number n of preimages of the root, the elementary graph $(\mathcal{G}_1, \Lambda_1)$ (note that elementary graphs form an ordered set by induction), and instead of the $(\mathcal{H}_v, \Lambda_v)$'s the graphs $(\mathcal{H}_{w^i_j}, \Lambda_{w^i_j})$, for w^i_j the special vertices of the elementary graph $(\mathcal{G}_1, \Lambda_1)$, taking into account the canonical order on $(\mathcal{G}_1, \Lambda_1)$. This concludes the proof. □

Bibliography

[1] S.S. Abhyankar; T.T. Moh. *Embedings of the line in the plane.* J.Reine Angew. Math, 276, 148-166, (1975).

[2] S.S. Abhyankar; A. Sathaye. *Uniqueness of plane embeddings of special curves.* Proc. Amer. Math. Soc 124 (1996) 1061-1069.

[3] S.S. Abhyankar; B. Singh. *Embeddings of certain curves in the affine plane.* Amer. J. Math. 100 no.1, 99-175, (1978).

[4] A. Bodin. *Classification of polynomials from $\mathbb{C}^2$ to $\mathbb{C}$ with one critical value.* Math. Z. 242, no. 2, 303–322, (2002).

[5] S.A. Broughton. *On the topology of polynomial hypersurfaces.* In P. Orlik editor, Proceedings of Simposia in Pure Mathematics 40, 167-178, Amer. Math. Soc. Providence RI, (1983).

[6] A.H. Durfee. *Five definitions of critical point at infinity.* Singularities, the Brieskorn anniversary volume. VI.Arnold, G.M.Greuel, J.H.M.Steenbrink editors. Progress in Mathematics 162, 345-360. Birkhauser 1998.

[7] J. Fernández de Bobadilla. *A new geometric proof of Jung's theorem on factorisation of automorphisms of $\mathbb{C}^2$.* To appear in Proc. Amer. Math. Soc.

[8] L. Fourrier. *Topologie d'un polynôme de deux variables complexes au voisinage de l'infini.* Ann.Inst.Fourier(Grenoble) 46 no.3, 645-687, (1996).

[9] G. Greuel; C. Hertling; G. Pfister. *Moduli spaces of semiquasihomogeneous singularities with fixed principal part.* J. Algebraic Geom. 6 no. 1, 169-199, (1997).

[10] S. Gusein-Zade; I. Luengo; A. Melle. *On the zeta function of a polynomial at infinity.* Bull. Sci. Math. 124:3, 213-224, (2000).

[11] G. Greuel; C.Lossen. *The geometry of families of singular curves.* New developments of singularity theory, D. Siersma; C.T.C. Wall; V. Zakalyukin editors. NATO Science series, II-vol 21., 159-192. (2001)

[12] G. Greuel; C. Lossen; E. Shustin. *Castelnuovo function, zero-dimensional ideals and singular plane curves.* J.Algebraic Geometry 9, 663-710, (2000).

[13] G. Greuel; C. Lossen; E. Shustin. *Plane curves of minimal degree with prescribed singularities.* Invent. Math. 133. no.3, 539-580, (1998).

[14] G. Greuel; G. Pfister. *Moduli for singularities.* Singularities (Lille 1991), 119-146, London Math. Soc. Lecture Note Ser. 201, Cambridge Univ. Press, Cambridge 1994.

[15] G. Greuel; G. Pfister. *Geometric quotients by unipotent group actions.* Proc. London Math. Soc. (3) 67 no. 1, 75-105, (1993).

[16] R. García Lopez; A. Nemethi. *On the monodromy at infinity of a polynomial map, I.* Compotitio Math. 100, 205-231, (1996).

[17] A. Grothendieck. *E.G.A. IV.* Publ. Math. I.H.E.S. 28, (1966).

[18] R. Hartshorne. *Algebraic Geometry.* Springer-Verlag. Graduate Texts in Mathematics 52, (1977).

[19] C. Hertling. *Frobenius manifolds and moduli spaces for hypersurface singularities.* Second part of his habilitationschrift. (2000).

[20] H.W.E. Jung. *Über ganze birationale Transformationen der Ebene.* J.Reine Angew. Math. 184, 1-15, (1942).

[21] S. Kleiman. *Multiple point formulae I: Iteration.* Acta. Math. 147, 13-49, (1981).

[22] S. Keel; S. Mori. *Quotients by groupoids.* Annals of Mathematics 145, 193-213, (1997).

[23] S. Lamy. *Une preuve géométrique du théorèm de Jung.* Enseign. Math. (2) 48, no. 3-4, 291-315, (2002).

[24] V.Ya. Lin; M.G. Zaidenberg. *An irreducible simply connected curve in $\mathbb{C}^2$ is equivalent to a quasihomogenous curve.* Soviet.Math.Dokl. 28, 200-204, (1983).

[25] I. Luengo; G. Pfister. *Normal forms and moduli spaces with semigroup* $\langle 2p, 2q, 2pq + d \rangle$. Compositio Math. 76 no 1-2, 247-264, (1990).

[26] D. Mumford. *Geometric Invariant Theory.* Springer-Verlag (1965).

[27] M. Nagata. *On automorphism group of k[x,y].* Lectures in Mathematics, Department of Mathematics, Kyoto University, 5. (1972)

[28] W.D. Neumann; P.G. Wightwick. *Algorithms for polynomials in two variables.* Combinatorial and computational algebra (Hong Kong, 1999), 219-235, Contemp.Math. 264. Amer. Math. Soc., Providence, RI, (2000).

[29] M. Oka. *Moduli space of affine curves of a given genus with one place at infinity.* Singularities, the Brieskorn anniversary volume. VI. Arnold, G. M. Greuel, J.H.M. Steenbrink editors. Progress in Mathematics 162, 409-434. Birkhauser (1998).

[30] J. Roé. *Varieties of clusters and Enriques diagrams.* Math.AG/018023.

[31] M. Alberich-Carraminana; J. Roe. *Enriques diagrams and adjacency of planar curve singularities.* Math.AG/0108223

[32] D. Siersma; J. Smeltink. *Classification of singularities at infinity of polynomials of degree 4 in two variables.* Georgian Math. Journal 7 no.1, 179-190, (2000) .

[33] D. Siersma; M. Tibar. *Singularities at infinity and their vanishing cycles, II. Monodromy.* Publications of the RIMS, Kyoto University 36, No 6, 659-679, (2000).

[34] V.A. Vassiliev. *Complements of Discriminants of Smooth Maps: Topology and Applications.* Revised ed. AMS, Transl. of Math. Monogr. 98, Providence RI (1994).

[35] V.A. Vassiliev. *How to calculate homology groups of spaces on nonsingular projective hypersurfaces in* $\mathbb{P}^n_{\mathbb{C}}$. Proc. Stelkov Math. Inst. 225, 121-140, (1999).

[36] V. Shpilrain; Jie-Tai Yu. *Embeddings of curves in the plane.* J. Algebra. 217, no.2, 668-678, (1999).

[37] V. Shpilrain; Jie-Tai Yu. *Non-extendable isomorphisms between affine varieties.* Math.AG/0110232

[38] H. Yoshihara. *Projective plane curves and the automorphism groups of their complements.* J. Math. Soc. Japan 37 no.1, 87-113, (1985).

Editorial Information

To be published in the *Memoirs*, a paper must be correct, new, nontrivial, and significant. Further, it must be well written and of interest to a substantial number of mathematicians. Piecemeal results, such as an inconclusive step toward an unproved major theorem or a minor variation on a known result, are in general not acceptable for publication. Papers appearing in *Memoirs* are generally longer than those appearing in *Transactions*, which shares the same editorial committee.

As of September 30, 2004, the backlog for this journal was approximately 5 volumes. This estimate is the result of dividing the number of manuscripts for this journal in the Providence office that have not yet gone to the printer on the above date by the average number of monographs per volume over the previous twelve months, reduced by the number of volumes published in four months (the time necessary for preparing a volume for the printer). (There are 6 volumes per year, each containing at least 4 numbers.)

A Consent to Publish and Copyright Agreement is required before a paper will be published in the *Memoirs*. After a paper is accepted for publication, the Providence office will send a Consent to Publish and Copyright Agreement to all authors of the paper. By submitting a paper to the *Memoirs*, authors certify that the results have not been submitted to nor are they under consideration for publication by another journal, conference proceedings, or similar publication.

Information for Authors

Memoirs are printed from camera copy fully prepared by the author. This means that the finished book will look exactly like the copy submitted.

The paper must contain a *descriptive title* and an *abstract* that summarizes the article in language suitable for workers in the general field (algebra, analysis, etc.). The *descriptive title* should be short, but informative; useless or vague phrases such as "some remarks about" or "concerning" should be avoided. The *abstract* should be at least one complete sentence, and at most 300 words. Included with the footnotes to the paper should be the 2000 *Mathematics Subject Classification* representing the primary and secondary subjects of the article. The classifications are accessible from `www.ams.org/msc/`. The list of classifications is also available in print starting with the 1999 annual index of *Mathematical Reviews*. The Mathematics Subject Classification footnote may be followed by a list of *key words and phrases* describing the subject matter of the article and taken from it. Journal abbreviations used in bibliographies are listed in the latest *Mathematical Reviews* annual index. The series abbreviations are also accessible from `www.ams.org/publications/`. To help in preparing and verifying references, the AMS offers MR Lookup, a Reference Tool for Linking, at `www.ams.org/mrlookup/`. When the manuscript is submitted, authors should supply the editor with electronic addresses if available. These will be printed after the postal address at the end of the article.

Electronically prepared manuscripts. The AMS encourages electronically prepared manuscripts, with a strong preference for $\mathcal{A}\mathcal{M}\mathcal{S}$-LaTeX. To this end, the Society has prepared $\mathcal{A}\mathcal{M}\mathcal{S}$-LaTeX author packages for each AMS publication. Author packages include instructions for preparing electronic manuscripts, the *AMS Author Handbook*, samples, and a style file that generates the particular design specifications of that publication series. Though $\mathcal{A}\mathcal{M}\mathcal{S}$-LaTeX is the highly preferred format of TeX, author packages are also available in $\mathcal{A}\mathcal{M}\mathcal{S}$-TeX.

Authors may retrieve an author package from e-MATH starting from `www.ams.org/tex/` or via FTP to `ftp.ams.org` (login as `anonymous`, enter username as password, and type `cd pub/author-info`). The *AMS Author Handbook* and the *Instruction Manual* are available in PDF format following the author packages link from `www.ams.org/tex/`. The author package can be obtained free of charge by sending email to `pub@ams.org` (Internet) or from the Publication Division, American Mathematical Society, 201 Charles St., Providence, RI 02904, USA. When requesting an author package, please specify $\mathcal{A}_{\mathcal{M}}\mathcal{S}$-LaTeX or $\mathcal{A}_{\mathcal{M}}\mathcal{S}$-TeX, Macintosh or IBM (3.5) format, and the publication in which your paper will appear. Please be sure to include your complete mailing address.

Sending electronic files. After acceptance, the source file(s) should be sent to the Providence office (this includes any TeX source file, any graphics files, and the DVI or PostScript file).

Before sending the source file, be sure you have proofread your paper carefully. The files you send must be the EXACT files used to generate the proof copy that was accepted for publication. For all publications, authors are required to send a printed copy of their paper, which exactly matches the copy approved for publication, along with any graphics that will appear in the paper.

TeX files may be submitted by email, FTP, or on diskette. The DVI file(s) and PostScript files should be submitted only by FTP or on diskette unless they are encoded properly to submit through email. (DVI files are binary and PostScript files tend to be very large.)

Electronically prepared manuscripts can be sent via email to `pub-submit@ams.org` (Internet). The subject line of the message should include the publication code to identify it as a Memoir. TeX source files, DVI files, and PostScript files can be transferred over the Internet by FTP to the Internet node `e-math.ams.org` (130.44.1.100).

Electronic graphics. Comprehensive instructions on preparing graphics are available at `www.ams.org/jourhtml/graphics.html`. A few of the major requirements are given here.

Submit files for graphics as EPS (Encapsulated PostScript) files. This includes graphics originated via a graphics application as well as scanned photographs or other computer-generated images. If this is not possible, TIFF files are acceptable as long as they can be opened in Adobe Photoshop or Illustrator. No matter what method was used to produce the graphic, it is necessary to provide a paper copy to the AMS.

Authors using graphics packages for the creation of electronic art should also avoid the use of any lines thinner than 0.5 points in width. Many graphics packages allow the user to specify a "hairline" for a very thin line. Hairlines often look acceptable when proofed on a typical laser printer. However, when produced on a high-resolution laser imagesetter, hairlines become nearly invisible and will be lost entirely in the final printing process.

Screens should be set to values between 15% and 85%. Screens which fall outside of this range are too light or too dark to print correctly. Variations of screens within a graphic should be no less than 10%.

Inquiries. Any inquiries concerning a paper that has been accepted for publication should be sent directly to the Electronic Prepress Department, American Mathematical Society, 201 Charles St., Providence, RI 02904, USA.

Titles in This Series

820 **Vadim A. Kaimanovich and Mikhail Lyubich,** Conformal and harmonic measures on laminations associated with rational maps, 2005

819 **F. Andreatta and E. Z. Goren,** Hilbert modular forms: mod p and p-adic aspects, 2005

818 **Tom De Medts,** An algebraic structure for Moufang quadrangles, 2005

817 **Javier Fernández de Bobadilla,** Moduli spaces of polynomials in two variables, 2005

816 **Francis Clarke,** Necessary conditions in dynamic optimization, 2005

815 **Martin Bendersky and Donald M. Davis,** v_1-periodic homotopy groups of $SO(n)$, 2004

814 **Johannes Huebschmann,** Kähler spaces, nilpotent orbits, and singular reduction, 2004

813 **Jeff Groah and Blake Temple,** Shock-wave solutions of the Einstein equations with perfect fluid sources: Existence and consistency by a locally inertial Glimm scheme, 2004

812 **Richard D. Canary and Darryl McCullough,** Homotopy equivalences of 3-manifolds and deformation theory of Kleinian groups, 2004

811 **Ottmar Loos and Erhard Neher,** Locally finite root systems, 2004

810 **W. N. Everitt and L. Markus,** Infinite dimensional complex symplectic spaces, 2004

809 **J. T. Cox, D. A. Dawson, and A. Greven,** Mutually catalytic super branching random walks: Large finite systems and renormalization analysis, 2004

808 **Hagen Meltzer,** Exceptional vector bundles, tilting sheaves and tilting complexes for weighted projective lines, 2004

807 **Carlos A. Cabrelli, Christopher Heil, and Ursula M. Molter,** Self-similarity and multiwavelets in higher dimensions, 2004

806 **Spiros A. Argyros and Andreas Tolias,** Methods in the theory of hereditarily indecomposable Banach spaces, 2004

805 **Philip L. Bowers and Kenneth Stephenson,** Uniformizing dessins and Belyĭ maps via circle packing, 2004

804 **A. Yu Ol'shanskii and M. V. Sapir,** The conjugacy problem and Higman embeddings, 2004

803 **Michael Field and Matthew Nicol,** Ergodic theory of equivariant diffeomorphisms: Markov partitions and stable ergodicity, 2004

802 **Martin W. Liebeck and Gary M. Seitz,** The maximal subgroups of positive dimension in exceptional algebraic groups, 2004

801 **Fabio Ancona and Andrea Marson,** Well-posedness for general 2×2 systems of conservation law, 2004

800 **V. Poénaru and C. Tanas,** Equivariant, almost-arborescent representation of open simply-connected 3-manifolds; A finiteness result, 2004

799 **Barry Mazur and Karl Rubin,** Kolyvagin systems, 2004

798 **Benoît Mselati,** Classification and probabilistic representation of the positive solutions of a semilinear elliptic equation, 2004

797 **Ola Bratteli, Palle E. T. Jorgensen, and Vasyl' Ostrovs'kyĭ,** Representation theory and numerical AF-invariants, 2004

796 **Marc A. Rieffel,** Gromov-Hausdorff distance for quantum metric spaces/Matrix algebras converge to the sphere for quantum Gromov-Hausdorff distance, 2004

795 **Adam Nyman,** Points on quantum projectivizations, 2004

794 **Kevin K. Ferland and L. Gaunce Lewis, Jr.,** The $RO(G)$-graded equivariant ordinary homology of G-cell complexes with even-dimensional cells for $G = \mathbb{Z}/p$, 2004

793 **Jindřich Zapletal,** Descriptive set theory and definable forcing, 2004

792 **Inmaculada Baldomá and Ernest Fontich,** Exponentially small splitting of invariant manifolds of parabolic points, 2004

791 **Eva A. Gallardo-Gutiérrez and Alfonso Montes-Rodríguez,** The role of the spectrum in the cyclic behavior of composition operators, 2004

790 **Thierry Lévy,** Yang-Mills measure on compact surfaces, 2003

789 **Helge Glöckner,** Positive definite functions on infinite-dimensional convex cones, 2003

788 **Robert Denk, Matthias Hieber, and Jan Prüss,** $\mathcal{R}$-boundedness, Fourier multipliers and problems of elliptic and parabolic type, 2003

787 **Michael Cwikel, Per G. Nilsson, and Gideon Schechtman,** Interpolation of weighted Banach lattices/A characterization of relatively decomposable Banach lattices, 2003

786 **Arnd Scheel,** Radially symmetric patterns of reaction-diffusion systems, 2003

785 **R. R. Bruner and J. P. C. Greenlees,** The connective K-theory of finite groups, 2003

784 **Desmond Sheiham,** Invariants of boundary link cobordism, 2003

783 **Ethan Akin, Mike Hurley, and Judy A. Kennedy,** Dynamics of topologically generic homeomorphisms, 2003

782 **Masaaki Furusawa and Joseph A. Shalika,** On central critical values of the degree four L-functions for GSp(4): The Fundamental Lemma, 2003

781 **Marcin Bownik,** Anisotropic Hardy spaces and wavelets, 2003

780 **S. Marmi and D. Sauzin,** Quasianalytic monogenic solutions of a cohomological equation, 2003

779 **Hansjörg Geiges,** h-principles and flexibility in geometry, 2003

778 **David B. Massey,** Numerical control over complex analytic singularities, 2003

777 **Robert Lauter,** Pseudodifferential analysis on conformally compact spaces, 2003

776 **U. Haagerup, H. P. Rosenthal, and F. A. Sukochev,** Banach embedding properties of non-commutative L^p-spaces, 2003

775 **P. Lochak, J.-P. Marco, and D. Sauzin,** On the splitting of invariant manifolds in multidimensional near-integrable Hamiltonian systems, 2003

774 **Kai A. Behrend,** Derived ℓ-adic categories for algebraic stacks, 2003

773 **Robert M. Guralnick, Peter Müller, and Jan Saxl,** The rational function analogue of a question of Schur and exceptionality of permutation representations, 2003

772 **Katrina Barron,** The moduli space of $N = 1$ superspheres with tubes and the sewing operation, 2003

771 **Shigenori Matsumoto,** Affine flows on 3-manifolds, 2003

770 **W. N. Everitt and L. Markus,** Elliptic partial differential operators and symplectic algebra, 2003

769 **Jie Wu,** Homotopy theory of the suspensions of the projective plane, 2003

768 **R. Höpfner and E. Löcherbach,** Limit theorems for null recurrent Markov processes, 2003

767 **Po Hu,** S-modules in the category of schemes, 2003

766 **Su Gao and Alexander S. Kechris,** On the classification of Polish metric spaces up to isometry, 2003

765 **Robert Bieri and Ross Geoghegan,** Connectivity properties of group actions on non-positively curved spaces, 2003

764 **J. Spandaw,** Noether-Lefschetz problems for degeneracy loci, 2003

For a complete list of titles in this series, visit the AMS Bookstore at **www.ams.org/bookstore/**.